Lecture Notes in Operations Research and Mathematical Systems

Economics, Computer Science, Information and Control

Edited by M. Beckmann, Providence and H. P. Künzi, Zürich

43

J.-A. Morales

Université Catholique de Louvain
Center for Operations Research and Econometrics.

Bayesian Full Information Structural Analysis

with an Application to the Study of the Belgian Beef Market

Springer-Verlag
Berlin · Heidelberg · New York 1971

AMS Subject Classifications (1970): 62 F 15, 62 P 20

ISBN 3-540-05417-0 Springer-Verlag Berlin Heidelberg New York
ISBN 0-387-05417-0 Springer-Verlag New York Heidelberg Berlin

Offsetdruck: Julius Beltz, Hemsbach/Bergstr.

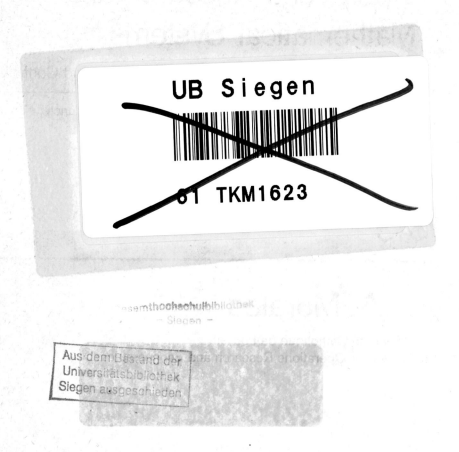

ACKNOWLEDGMENTS

There are many people to whom I am indebted. First of all, I am very grateful to my teacher Jacques H. Drèze. He suggested the topic of this dissertation and he has offered generous advise and constant encouragement. It is beyond doubt that this dissertation could never have been written without Professor Drèze's numerous suggestions ; many points discussed in the text below are to be considered as a continuation of the developments laid out by him in several papers since 1962.

Professors Anton Barten, Louis Phlips, Peter Schönfeld, Emiel van Broekhoven and Mr. Michel Hesbois have commented on a preliminary version. Their remarks have helped very much to improve the presentation. Bernard Calicis has kindly allowed me to use his data on the Belgian beef market, and I have also benefited from many conversations with him. My friends and colleagues Michel Mouchart and Jean-François Richard have given useful advise and constructive criticism at all stages of the dissertation. Without the numerical integration computer programs written by Jean-François Richard, the realization of the empirical illustration would have been very difficult. Cécile Morales-Belpaire has helped greatly in the painful job of proof reading ; furthermore I have enjoyed her comments on the Belgian beef market.

Over the years that this dissertation has been developed I have been supported by an assistantship at the Institut des Sciences Economiques and an association to the Center for Operations Research and Econometrics, both at the Université Catholique de Louvain. Acknowledgment is also due to the Louvain Computing Center where all the extensive calculations that were needed have been made.

Contents

Introduction

1. Statement of the problem.[1]

Bayes' theorem provides a very powerful tool for statistical inference, especially when pooling information from different sources is appropriate. Thus, prior information resulting from economic theory and/or from previous (real or hypothetical) samples can be combined with the information embodied in new observations ; and this operation can be performed formally, within a rigorous mathematical framework.

To introduce the Bayesian analysis of the simultaneous equations model, we shall base our presentation in the very convenient exposition given by Drèze in his presidential adress to the Second World Congress of the Econometric Society.[2]

The Bayesian method in statistics is usually presented as follows : Consider the joint probability density function $f(x,\theta)$ defined on the product space $X \times \boldsymbol{\Theta}$, where $X = \{x\}$ denotes the sample space, and $\boldsymbol{\Theta} = \{\theta\}$ denotes the parameter space. If we decompose the joint density $f(x,\theta)$ in a conditional density $f(x/\theta)$ and a marginal

---

[1] The beginning of this section reviews some very well known propositions of Bayesian analysis. Those who are familiar with the subject can skip this part, and start with p.5.

[2] J.H.Drèze. "Econometrics and Decision Theory". Presidential adress delivered at the Second World Congress of the Econometric Society. Available also as CORE Discussion Paper N° 7040, Louvain.

density $f(\theta)$ it follows from Bayes' theorem that[1]

$$f(\theta/x) = \frac{f(\theta)f(x/\theta)}{\int_{\Theta} f(\theta)f(x/\theta)d\theta}$$

Notice that Bayes' theorem uses two inputs under the form of density functions. The first input is the marginal density function $f(\theta)$ defined "prior" to the observation. The second input is the density function of the observations conditional to a given value of θ. This function $f(x/\theta)$ viewed as a function of θ, for a given sample, is the likelihood function of classical statistics. The output of Bayes' theorem is the posterior density function $f(\theta/x)$.[2] Probability judgements on θ, and of future values of x will be made according to this posterior density function.

The first step consists in choosing a density that represents satisfactorily our prior knowledge. Numerical applications in Bayesian Econometrics have relied principally on "non-informative" priors defined by Jeffreys' invariance principle as in much of the work of Zellner[3], or on families of natural conjugate priors as introduced by Pratt, Raiffa, Schlaiffer and their associates.[4]

---

[1] Here and in the text we use the symbol f to denote a density function in general. The argument of the function identifies the particular distribution being considered. However, when there is a risk of ambiguity, a particular symbol will identify the function.

[2] We find convenient in the text to denote by $f^{o}(\theta)$ the prior density function, by $L(\theta/x)$ the likelihood function and by $f^{*}(\theta/x)$ the posterior density.

[3] For applications to the Bayesian analysis of systems of equations, see e.g. Tiao and Zellner (29), Zellner (35).

[4] A standard reference to this approach is (22). Application to systems of equations can be found in Ando and Kaufman (2).

For the most common data generating processes, the "non-informative" priors of the type mentioned above take the form of an uniform density on the range of θ, if the range is finite or if it goes from $-\infty$ to $+\infty$, and of an uniform density on the logarithm of θ, if θ goes from 0 to $+\infty$. For example, if data are generated by a Normal process with mean μ, and precision $h = 1/\sigma^2$, the prior for μ, say $f_1(\mu)$ would be $f_1(\mu) \propto 1$, and the prior for h, say $f_2(h)$ will be $f_2(h) \propto h^{-1}$. The joint density $f(\mu,h)$ would be usually presented as $f(\mu,h) = f_1(\mu)f_2(h) \propto h^{-1}$.

Families of natural conjugate prior densities have also been defined for the most common data generating processes (like Binomial, Poisson, Normal Sampling). For those processes the density of the observations $f(x/\theta)$ can be factored in two functions $k_1(u(x)/\theta)$ and $(k_2(x))$, with $u(x)$ a sufficient statistic of fixed dimensionality summarizing the information of the observations about θ ; $k_1(u(x)/\theta)$ is labeled a "kernel" function and $k_2(x)$ is a residual function. Interchanging variable and parameter in $k_1(u(x)/\theta)$ we obtain the kernel of a natural conjugate prior. If the integral $N(u) = \int_\Theta k(\theta/u(x)d\theta$ exists, then the function $f(\theta/u(x)) = \frac{1}{N(u)} k(\theta/u(x))$ is a proper natural conjugate prior density function with parameter $u(x)$.

The families of natural conjugate prior densities on the process parameters have the following useful properties : 1) They are closed under sampling, i.e. the posterior density arising from a prior density in the family, is also a member of the family. 2) The parameters of the posterior distributions can be readily computed as functions of the corresponding parameters of the prior density function and of the sufficient statistics summarizing the sample information. With these families "the kernel of the prior density combines with the sample kernel in exactly the same way as two sample kernels combine."[1]

When one wishes to incorporate in the analysis prior information which is not equivalent to what might have been learned from a hypothetical previous sample generated by the same process, it is sometimes possible to extend the natural conjugate family by introducing additional parameters.

Concerning the likelihood function $f(x/\theta)$, it is important to note the following definitions. If for two parameter points θ_1 and θ_2 in Θ, and for all x, $f(x/\theta_1) = f(x/\theta_2)$, then θ_1 and θ_2 are said to be observationally equivalents. Therefore the definition of identifiability in classical Econometrics follows : "a parameter

---

[1] Raiffa and Schlaiffer (22) p. 49

point θ^* in Θ is said to be identified if there is no other θ in Θ which is observationally equivalent."[1]

The principles stated above can be applied very naturally to the analysis of the model of simultaneous equations in structural form. Prior information plays a role of paramount importance in the analysis of simultaneous equations in view of the identification problem.

In a traditional context prior information takes generally the form of exact constraints on some parameters of the model. But, a more general and flexible treatment can be gained by introducing stochastic constraints on all or a subset of the structural parameters, instead of deterministic constraints.

Consider the structural system $By+\Gamma z = u$, with reduced form $y = \Pi z+v$. Assume that v is distributed Normally with mean zero and covariance matrix Ω. Let $\Sigma = B\Omega B'$ denote the covariance matrix of the structural form disturbances. The structural parameters are thus (B,Γ,Σ) and the reduced form parameters are (Π,Ω).[2] From a sampling theory viewpoint, an identification problem arises because the likelihood function remains stationary when written

[1] See e.g. Rothenberg (26). The concept of identification from a Bayesian viewpoint is different, as is shown below. See also Rothenberg (25).

[2] The specification of the model is considered more formally in Part I, section 1.1. in the text.

either as a function of the reduced form parameters or as a function of the structural parameters. Since the correspondence between structural and reduced form parameters is many-to-one, there is an identification problem. That problem is solved by restricting the space of the structural parameters in such a way that a one-to-one correspondence emerges.

In a Bayesian context, one can define a prior density function $f^o(B,\Gamma,\Sigma)$ on the unrestricted or restricted space of structural parameters, and combine it with the likelihood function $L(B,\Gamma,\Sigma/Y,Z)$. Using Bayes' theorem, the posterior density is :

$$f^*(B,\Gamma,\Sigma/Y,Z) \propto f^o(B,\Gamma,\Sigma)L(B,\Gamma,\Sigma/Y,Z).$$

This operation is well defined, whether the model is identified from a sampling theory viewpoint or not. But, we have the important result of Drèze concerning the problem of identification.[1] "Conditionally on a set of values for the reduced form parameters, the prior and the posterior densities are identical : $f(B/\Pi,\Omega,Y,Z) \underset{Y,Z}{\equiv} f^o(B/\Pi,\Omega)$." This identity reveals that the "observations are informative about the reduced form parameters alone . Exact identifying restrictions correspond to the special case where $f^o(B/\Pi,\Omega)$ has all its mass concentrated at a single point."

[1] See Drèze (10), (11). The problems related to the identification problem are considered more formally in Part I, section 1.2.

In the study of systems of equations in structural form, two approaches have been followed as well in classical Econometrics as in Bayesian Econometrics. The first one is labeled full information because prior information and estimation (or computation of the posterior parameters) bear on the whole system. This way of proceeding, as it is well known, is more efficient : each equation of the system benefits from the information on all equations.[1] The second approach disregards some of the a priori information, uses only the prior information on the parameters of a single equation, and likewise proceeds with the estimation (or the computation of the posterior parameters) for a single equation.

A Bayesian treatment of the latter approach can be found in Drèze[2] or Zellner[3]. This dissertation is concerned uniquely with the Full Information procedure, i.e., with the problem of incorporating prior information on all (or a subset) of the parameters of the whole model, and with the problem of obtaining the posterior distributions of those parameters.

If we want to do Bayesian inference on the <u>unrestricted</u> parameter space, we must face the difficulty of defining a satisfactory

---

[1] By the same token, this approach adds sensitivity to specification errors.

[2] Drèze (11)

[3] Zellner (35)

prior density. Indeed, if we use a "non-informative" prior such that $f^o(B/\Pi,\Omega) = f^o(B) \propto k$, the posterior density will be marginaly uniform over the complete space of the B's and no inference is possible. Natural conjugate priors will not work either since in this approach we reason implicitely in terms of previous samples, and we know that a sample is non-informative about the structural parameters.

Even if the prior density $f^o(B/\Pi,\Omega)$ were uniform, it would be possible to derive a natural conjugate prior on Γ conditional on B. But even so, the natural conjugate approach appears as unsatisfactory since there remains the problem of Rothenberg. Rothenberg[1] pointed out for the reduced form model that the covariance matrix of the regression coefficients embodied an unwanted restriction, namely, the covariance matrix of the regression coefficients of the ith equation is equal to the covariance matrix of the regression coefficients of the jth equation ($j \neq i$) up to a scalar multiple.

Indeed, given the assumptions above on the distribution of the reduced form disturbances, the natural conjugate prior for the reduced form model will read :[2]

---

[1] Rothenberg (24)

[2] A more complete formulation can be found in Part I, section 1.2. in the text

$$f^{o}(\Pi,\Omega^{-1}) \propto |\Omega^{-1}|^{\frac{\nu}{2}} \exp - \frac{1}{2} \operatorname{tr} \Omega^{-1}\{(\Pi-\Pi^{o})M^{o}_{zz}(\Pi-\Pi^{o})'+S^{o}\}$$

The joint density $f^{o}(\Pi,\Omega^{-1})$ appears as a Normal-Wishart density on Π, Ω^{-1}. The parameters of this density are given by Π^{o}, M^{o}_{zz}, S^{o}, and the dimensions of Π, Ω^{-1}.[1]

The conditional density $f^{o}(\Pi/\Omega^{-1})$ is a Normal density function with mean Π^{o} and covariance matrix $V(\Pi) = (\Omega \otimes M^{o\,-1}_{zz})$. From the form of this covariance matrix is seen that the covariance matrix of the regression coefficients of the ith equation, namely $\omega_{ii}M^{o\,-1}_{zz}$ is a scalar multiple of the covariance matrix of the regression coefficients of the jth equation, namely $\omega_{jj}M^{o\,-1}_{zz}$.

This property of the covariance matrix of Π will be preserved when passing to the conditional covariance matrix of Γ since

$$\begin{aligned} V(\Gamma|B) &= V(-B\Pi) \\ &= (B \otimes I)V(\Pi)(B' \otimes I) \\ &= (B\Omega B' \otimes M^{o\,-1}_{zz}) \\ &= (\Sigma \otimes M^{o\,-1}_{zz}) \quad . \end{aligned}$$

This proportionality of the variances is indeed a severe restriction. In the special case, that arises often in practice, where

---

[1]This presentation has to be considered as introductory, the parameter set of $f^{o}(\Pi,\Omega^{-1})$ is defined more precisely in the text.

we have informations from different sources, one for each equation, it is difficult to incorporate them in a natural conjugate approach. Drèze[1] has proposed instead, for the analysis in structural form, an "extended" natural conjugate density function that overcomes Rothenberg's objection by attempting to generalize the covariance matrix so as to allow non-proportional covariance matrices of different equations.

Results on the use of this extended natural conjugate density have been obtained by Drèze and the author.[2] Additional results are reported here.

2. Review of the contents.

In this dissertation we develop some theoretical aspects connected with the extended natural conjugate prior density and provide an empirical illustration.

Part I is concerned with the theoretical aspects. Section 1 introduces the statistical model, and reviews the known results concerning the problem of identification in a Bayesian context. The performance of non-informative priors and natural conjugate priors is also re-examined there. Section 2 proposes the extended natural conjugate density along with a proper interpretation and some

---

[1] Drèze (11)

[2] Drèze and Morales (12)

properties[1]. That section contains lengthy proofs on sufficient conditions for the extended natural conjugate prior density to be a proper density ; these proofs are not necessary for understanding the rest of the book. In section 3, we derive the posterior distributions of the structural parameters, and we show that the proposed "extended natural conjugate prior density" belongs to a closed family of prior densities. In this section, we analyse also the difficulties of computing analytically the moments of the posterior distributions of all the parameters and we propose instead, an alternate solution by showing that standard algorithms used to obtain full information maximum likelihood estimates in classical analysis, can be employed with slight modifications to compute the modal values of the posterior distribution.

Part II is concerned with an empirical application of the results of Part I. This empirical application bears on a two-equations model of supply and demand for beef. This model is a modified version of a larger model proposed by Calicis to study the Belgian beef market.[2]

---

[1] Results for the special case of a two-equations model are more complete than those for the general model.

[2] Calicis (5).

Section 1 states the model and specifies the prior distributions. In section 2, the posterior analysis is done. In this section we compute by numerical integration methods, the moments of the posterior distributions, and we also give intervals containing the posterior modes. In this section we give also the main conclusions of the empirical illustration.

Part I

Bayesian Full Information Analysis of the Simultaneous Equations Model.

1. A review of the problem of identification in a Bayesian approach and the specification of prior density functions.

1.1. The statistical model and notation.

Let observations be generated by the following system of structural equations :

$$(1.1) \qquad By(t) + \Gamma z(t) = u(t)$$

where $y(t)$ is an m dimensional vector of endogenous variables, $z(t)$ is an n-dimensional vector of truly exogenous variables, B is an m x m non-singular matrix of coefficients with diagonal elements $\beta_{ii} = 1$, $i = 1,2,\ldots,m$, Γ is an m x n matrix of coefficients and $u(t)$ is an m-dimensional vector of unobservable random disturbances with a non-degenerate joint distribution[1].

If T observations are available, the system 1.1. can be written as :

$$(1.2.) \qquad BY' + \Gamma Z' = U'$$

where $Y = (y_1,y_2,\ldots y_m)$ and $U = (u_1,u_2,\ldots u_m)$ are T x m matrices, and $Z = (z_1,z_2\ldots z_n)$ is a T x n matrix. The corresponding reduced form is :

$$(1.3.) \qquad Y' = -B^{-1}\Gamma Z' + B^{-1}U' \underset{\text{def}}{=} \Pi Z' + V'.$$

[1] Notice that there are no identities.

For given Z and Π, the observations Y are assumed to be generated by a random process entirely defined by the joint distribution of the reduced form disturbances. We shall assume that the rows of V are mutually independent Normal m-dimensional vectors with mean zero and variance-covariance matrix Ω. Thus the column expansion of V is distributed Normally with zero mean and variance-covariance matrix $(\Omega \otimes I)$. The density of U obtain d through an integrand transformation is also Normal with zero mean and variance-covariance matrix $(\Sigma \otimes I)$; with $\Sigma \underset{\text{def}}{=} B\Omega B'$.

The likelihood of the observations Y obtained by using a transformation from the U-space to the Y-space is :

$$(1.4.)\quad L(B,\Gamma,\Sigma/Y,Z) \propto ||B||^{T}|\Sigma^{-1}|^{\frac{T}{2}}\exp - \tfrac{1}{2}\mathrm{tr}\ \Sigma^{-1}(B\Gamma)\begin{bmatrix} Y'Y & Y'Z \\ Z'Y & Z'Z \end{bmatrix}\begin{bmatrix} B' \\ \Gamma' \end{bmatrix}.$$

The representations of the likelihood function as given below will be needed to derive posterior distributions.[1] The usefulness of these representations stems from the fact that they show explicitely the normalization rule. Also, these representations make easy the introduction of exact restrictions if we want to depart from our, according to the classical viewpoint, completely unidentified model .

Consider a single equation of the system 1.1.

---

[1]See section 3.1

$$(1.5.)\qquad \begin{aligned} y_i &= -\sum_{j\neq i} y_j\beta_{ij} - \sum_k z_k\gamma_{ik} + u_i \\ &\underset{\text{def}}{=} Y_i b_i + Zc_i + u_i \end{aligned}$$

$$i = 1,2,\ldots,m,$$

where $Y_i = (y_1,y_2,\ldots,y_{i-1},y_{i+1},\ldots,y_m)$ is a $(T\ x(m-1))$-dimensional matrix ; $Z = (z_1,z_2,\ldots,z_k,\ldots,z_n)$ is a $(T\ x\ n)$-dimensional matrix as before, b_i is an $(m-1)$ dimensional vector of coefficients of the endogenous variables in the right hand of 1.5 affected with a minus sign, and c_i is an n-dimensional vector of coefficients of the exogenous variables affected with a minus sign.

Defining furthermore $X_i = (Y_i:Z)$, and $\delta_i' = (b_i':c_i')$ and putting together the m-equations 1.5., we can write

$$(1.6.)\qquad \begin{pmatrix} y_1 \\ y_2 \\ \vdots \\ y_m \end{pmatrix} = \begin{pmatrix} X_1 & 0 & \cdots & 0 \\ 0 & X_2 & & \\ \vdots & & \ddots & \\ 0 & & & X_m \end{pmatrix} \begin{pmatrix} \delta_1 \\ \delta_2 \\ \vdots \\ \delta_m \end{pmatrix} + \begin{pmatrix} u_1 \\ u_2 \\ \\ u_m \end{pmatrix}$$

or still

$$y \underset{\text{def}}{=} X\delta + u$$

where y and u are mT-dimensional vectors, X is an $(mT\ x\ m(n+m-1))$ matrix and δ an $m(n+m-1)$-dimensional vector.

Thus $-\delta_i'$ is the ith row of $(B:\Gamma)$ with the ith element deleted, and X_i is the matrix $(Y : Z)$ with the ith column deleted.

The following alternate notation of our model in 1.6. will also be useful for deriving posterior distributions in section 3 below :

$$(1.7.)\qquad Y = (X_1 X_2 \ldots X_m) \begin{bmatrix} \delta_1 & 0 & \ldots & 0 \\ 0 & \delta_2 & & \\ \vdots & & & \\ 0 & \ldots & \ldots & \delta_m \end{bmatrix} + U$$

$$Y = \Xi \, \Delta' + U$$

where Y is the T x m matrix of endogenous variables, Ξ is a T x m(m + n-1) matrix, Δ is a m x m(m + n-1) matrix defined as :

$$\Delta = \begin{bmatrix} \delta_1' & \ldots & \ldots & 0 \\ 0 & \delta_2' & & \\ \vdots & & & \\ 0 & \ldots & \ldots & \delta_m' \end{bmatrix} ,$$

and U is the T x m matrix of structural disturbances[1].

In these new notations, the likelihood function in 1.4. can be written as :

$$(1.8.)\qquad L(\delta,\Sigma/Y,Z) \propto \|B\|^T |\Sigma^{-1}|^{\frac{T}{2}} \exp - \tfrac{1}{2}(y-X\delta)'(\Sigma^{-1} \otimes I)(y-X\delta)$$

[1]As mentioned above, we would like to start with a model that is completely unidentified from a classical viewpoint.

If we want to treat a model that embodies exact exclusion restrictions, only the non-restricted coefficients should be retained in δ. Also the columns of (Y:Z) corresponding to the exclusion restrictions of the ith equation, i=1,2,...,m, should be deleted to obtain each X_i in 1.6. If enough restrictions are placed on the model to identify it, according to a classical viewpoint, then the notation in 1.6. corresponds to the notations given by Zellner and Theil (33), and Rothenberg and Leenders (23).

or also as :

$$(1.9.)\quad L(\delta,\Sigma/Y,Z) \propto ||B||^{T}|\Sigma^{-1}|^{\frac{T}{2}}\exp -\frac{1}{2}\,\mathrm{tr}\,\Sigma^{-1}(Y - \Xi\Delta')'(Y - \Xi\Delta).$$

1.2. <u>The problem of identification in a Bayesian context and the choice of prior distributions.</u>

The problem of identification for the simultaneous equation model has been presented classically in the following terms.

The identifiability of the structural parameters is established on the basis of the reduced form parameters. The likelihood of the observations Y obtained by a linear transformation from the density of V is written as :

$$(1.10.)\quad L(\Pi,\Omega/Y,Z) \propto |\Omega|^{-\frac{T}{2}}\exp -\frac{1}{2}\,\mathrm{tr}\,\Omega^{-1}[Y-Z\Pi']'[Y-Z\Pi'] .$$

As there exists a set of functions mapping the parameters' space (B,Γ,Σ) onto (Π,Ω), the likelihood in 1.4. is stationary over all values of (B,Γ,Σ) that satisfy the relations $-B^{-1}\Gamma = \hat{\Pi}$, and $B^{-1}\Sigma B'^{-1} = \hat{\Omega}$ for any two matrices $\hat{\Pi}$, and $\hat{\Omega}$. The problem of identification becomes then in this context, the problem of finding a <u>unique</u> set $(\hat{B},\hat{\Gamma},\hat{\Sigma})$ that satisfies these relations.

In a classical context, one solves this problem of identification by imposing on the structural parameters, a set of continuously differentiable constraint equations $g_k(B,\Gamma,\Sigma) = 0 \quad k=1,2,\ldots,K$ that associate with every point in the space of the reduced form parameters (Π,Ω) at most one point in the space of structural parameters (B,Γ,Σ).[1]

In a Bayesian approach, our prior information on the parameters takes the form of a joint (subjective) probability function defined on the space (or a subspace) of the parameters (B,Γ,Σ), say $f^{\circ}(B,\Gamma,\Sigma)$; we then combine this density with the likelihood function $L(B,\Gamma,\Sigma/Y,Z)$ by application of Bayes' theorem and the joint posterior density is therefore defined by :

$$(1.11) \quad f^{*}(B,\Gamma,\Sigma/Y,Z) = \frac{f^{\circ}(B,\Gamma,\Sigma)L(B,\Gamma,\Sigma/Y,Z)}{\int_{\Theta} f^{\circ}(B,\Gamma,\Sigma)L(B,\Gamma,\Sigma/Y,Z)dBd\Gamma d\Sigma} \propto f^{\circ}(B,\Gamma,\Sigma)L(B,\Gamma,\Sigma/Y,Z)$$

where Θ denotes the parameter space.

In many applications, we shall deal with a mixed approach to express our prior information, i.e. the prior information takes the form of exact constraints on some parameters (or groups of parameters) and of stochastic constraints on some other parameters (or groups of parameters). In fact, in this mixed approach, we

[1]This statement can be made more precise by stating the following result due to Rothenberg (26) : a necessary condition for the identification of (B,Γ,Σ) is that there be at least m^2 (including the normalisation identity) independent restrictions g_k.

define first the support of the density using the exact constraints, and then we define the density function on this support. Thus, the Bayesian analysis can be more or less close[1] to the classical one ; but the difference in nature between those two types of analyses remains ; since one of them leads to a (posterior) probability distribution of the parameters, and the other to point estimates (and their variances) for the parameters of the reduced form and of the structure, if identified.

The concept of identification in a Bayesian context is not altogether clear. We shall adopt the view of considering a structure "identified", if the <u>posterior</u> density of the parameters of the model is not "flat" on a subspace of the parameter space.[2] This point of view may not be entirely satisfactory, but a discussion on the concept of identification is beyond the scope of this dissertation.[3]

In a model for which we define a prior joint probability distribution on all parameters, there arises a problem related to the

---

[1] In a limiting case, with enough identifying exact restrictions, the Bayesian treatment can rely on the very same type of prior information as the classical treatment.

[2] Remark that we require the (posterior) density function not to be flat, without any reference to the integrability of the function.

[3] A more complete exposition on the problem of defining a satisfactory concept of identification can be found in Rothenberg (25), and Zellner (35). See also Drèze (10).

concept of identification . This problem has been formulated by Drèze[1] in the following way :

From the prior density function $f^{o}(B,\Gamma,\Sigma)$ defined on the unrestricted space of the structural parameters, we can obtain the density function $h^{o}(B,\Pi,\Omega)$ through an integrand transformation ;

$$\begin{aligned} h^{o}(B,\Pi,\Omega) &= f^{o}(B,-B\Pi,B\Omega B')\ ||B||^{n+m+1} \\ (1.12) \qquad &= h_1^{o}(B/\Pi,\Omega)\ h_2^{o}(\Pi,\Omega)\ . \end{aligned}$$

Similarly the posterior distribution $h^{*}(B,\Pi,\Omega)$ can be written as :

$$(1.13) \qquad h^{*}(B,\Pi,\Omega/Y,Z) = f^{*}(B,-B\Pi,B\Omega B'/Y,Z)\ ||B||^{n+m+1}.$$

The right hand term in 1.13 is equal to :

$$\frac{f^{o}(B,-B\Pi,B\Omega B')||B||^{n+m+1}\ L(\Pi,\Omega/Y,Z)}{\int_{B,\Pi,\Omega} f^{o}(B,-B\Pi,B\Omega B')||B||^{n+m+1}L(\Pi,\Omega/Y,Z)dBd\Pi d\Omega} =$$

$$h_1^{o}(B/\Pi,\Omega)\ \frac{h_2^{o}(\Pi,\Omega).L(\Pi,\Omega/Y,Z)}{\int_{\Pi,\Omega} h_2^{o}(\Pi,\Omega).L(\Pi,\Omega/Y,Z)d\Pi d\Omega}\ ,$$

and therefore :

$$(1.14) \qquad h^{*}(B,\Pi,\Omega/Y,Z) = h_1^{o}(B/\Pi,\Omega).h_2^{*}(\Pi,\Omega/Y,Z)$$

---

[1]Drèze (10),(11).

Thus we have the following theorem :

Theorem 1.1. (Drèze). Conditionally on Π and Ω, the posterior density of B is identical with the prior density, and thus remains unaffected by the observations (Y,Z).

An alternative statement of the theorem would be : "Conditionally on Π, Ω, the joint posterior density of (B,Γ,Σ) is identical with the prior density and remains unaffected by the observations (Y,Z) ; indeed Γ and Σ are obtained as unique functions of B,Π,Ω."
Bayesian methods in statistics have used mostly two kinds of priors, non-informative priors, and natural conjugate priors. If the statistical inference must bear on all parameters, non-informative priors are useless : the posterior density is then proportional to the likelihood $L(B,\Gamma,\Sigma/Y,Z)$[1] and the structure remains underidentified.

There is a difficulty of the same nature in using natural conjugate prior densities. The natural conjugate approach reasons implicitely in terms of previous samples, or of fictitious samples that are generated by the same process as the data. But the identification problem remains ; the parameters of the prior distribution cannot all be identified by the "previous" data anymore than by the new

[1] It is possible of course, to do Bayesian inference with non-informative priors when the model is identified by prior exact restrictions. But are we still non-informative?

data, in the absence of identifying restrictions. One way of seeing this is to think of the natural conjugate prior distribution as the posterior distribution arising from a non-informative prior and the likelihood of an underidentified structure.

This point deserves some attention. Consider as a prior density function the posterior density resulting from a non-informative prior and a likelihood function similar to 1.4 i.e.

$$(1.15.)\ f^{o}(B,\Gamma,\Sigma^{-1}) \propto f_1^{o}(B,\Gamma,\Sigma^{-1})\,||B||^{\theta}\,|\Sigma^{-1}|^{\frac{\theta}{2}} \exp -\tfrac{1}{2}\mathrm{tr}\ \Sigma^{-1}(B\Gamma)\begin{bmatrix} M^{o}_{yy} & M^{o}_{yz} \\ M^{o}_{zy} & M^{o}_{zz} \end{bmatrix}\begin{bmatrix} B' \\ \Gamma' \end{bmatrix}$$

The notation θ, $M^{o} = \begin{bmatrix} M^{o}_{yy} & M^{o}_{yz} \\ M^{o}_{zy} & M^{o}_{zz} \end{bmatrix}$ has been employed to denote a "previous sample". To gain flexibility, we do not require that θ be a positive integer.

Remark furthermore that the prior density in 1.15 bears on the elements of Σ^{-1} rather than Σ. This is done for convenience.

There remains to define a non-informative prior $f_1^{o}(B,\Gamma,\Sigma^{-1})$. We shall do this by first defining a non-informative prior on (B,Π,Ω^{-1}) of the form

$$g_1^{o}(B,\Pi,\Omega^{-1}) \propto |\Omega^{-1}|^{-\frac{m+1}{2}} \quad ;$$

then, $f_1^o(B,\Gamma,\Sigma^{-1})$ will be obtained through an integrand transformation from (B,Π,Ω^{-1}) to (B,Γ,Σ^{-1}), so that

$$f_1^o(B,\Gamma,\Sigma^{-1}) \propto |B'\Sigma^{-1}B|^{-\frac{m+1}{2}} |J|$$

$$(1.16.) \qquad \propto |\Sigma^{-1}|^{-\frac{m+1}{2}} ||B||^n$$

where $|J| = ||B||^{-n+m+1}$ is the Jacobian of the transformation from (B,Π,Ω^{-1}) to (B,Γ,Σ^{-1}).

Inserting 1.16 in 1.15 and integrating out Σ^{-1} we obtain the marginal density of (B,Γ).[1]

$$(1.17.) \quad f^o(B,\Gamma) \propto ||B||^{\theta-n} \left| \begin{bmatrix} B & \Gamma \end{bmatrix} \begin{bmatrix} M^o_{yy} & M^o_{zy} \\ M^o_{yz} & M^o_{zz} \end{bmatrix} \begin{bmatrix} B' \\ \Gamma' \end{bmatrix} \right|^{-\frac{\theta}{2}} =$$

$$||B||^{\theta-n} |(\Gamma+BM^o_{yz}M^{o-1}_{zz})M^o_{zz}(\Gamma+BM^o_{yz}M^{o-1}_{zz})' + BM^o_{yy.z}B'|^{-\frac{\theta}{2}}$$

where $M^o_{yy.z} = M^o_{yy} - M^o_{yz} M^{o-1}_{zz} M^o_{yz}$.

If $\theta > m+n-1$, the expression

$$|(\Gamma+BM^o_{yz}M^{o-1}_{zz})M^o_{zz}(\Gamma+BM^o_{yz}M^{o-1}_{zz})' + BM^o_{yy.z}B'|^{-\frac{\theta}{2}} ,$$

is the kernel of a proper matric variate-t-density on Γ,[2] with parameters $(BM^o_{yy.z}B', M^o_{zz}, BM^o_{yz}M^{-1}_{zz}, \theta)$.

---

[1] The integration with respect to Σ^{-1} can be performed, provided that $\theta > m-1$, since the conditional density $f^o(\Sigma^{-1}|B,\Gamma)$ is a Wishart-density with parameters $(\theta,(B:\Gamma)M^o(B:\Gamma)')$. For properties of the Wishart-density function, see Appendix Section A.

[2] For the definition of a matric variate-t-density and its properties, see Appendix Section B.

Thus, if we write 1.17 as :

$$(1.18)\quad f^{o}(B,\Gamma)\propto ||B||^{\theta-n}|BM^{o}_{yy.z}B'|^{-\frac{\theta-n}{2}}\;|BM^{o}_{yy.z}B'|^{\frac{\theta-n}{2}}\;.$$

$$|(\Gamma+BM^{o}_{yz}M^{o-1}_{zz})M^{o}_{zz}(\Gamma+BM^{o}_{yz}M^{o-1}_{zz})' + BM^{o}_{yy.z}B'|^{-\frac{\theta}{2}},$$

we can decompose the function $f^{o}(B,\Gamma)$ in 1.18 in

$f^{o}(B)\;.\;f^{o}(\Gamma|B)$ with :

$$(1.19)\quad f^{o}(B)\propto ||B||^{\theta-n}|BM^{o}_{yy.z}B'|^{-\frac{\theta-n}{2}}$$

$$\propto k$$

and

$$(1.20)\quad f^{o}(\Gamma|B)\propto|BM^{o}_{yy.z}B'|^{\frac{\theta-n}{2}}|(\Gamma+BM^{o}_{yz}M^{o-1}_{zz})M^{o}_{zz}(\Gamma+BM^{o}_{yz}M^{o-1}_{zz})'+BM^{o}_{yy.z}B'|^{-\frac{\theta}{2}},$$

where k denotes a constant.

The marginal density of B is thus "flat", reflecting indeed the identification problem. The conditional density of Γ given B in 1.20, has moments

$$(1.21)\quad \begin{aligned} E[\Gamma|B] &= -BM^{o}_{yz}M^{o-1}_{zz} \qquad \text{and} \\ V[\Gamma|B] &= \frac{1}{\theta-m-n-1}\,(BM^{o}_{yy.z}B'\otimes M^{o-1}) \quad \text{if} \quad \theta > m+n+1. \end{aligned}$$

The marginal moments of Γ do not exist since they depend on the moments of B. This is also a consequence of the identification problem.

The results in 1.19 and in 1.20 are not surprising. In fact they result from a particular application of the theorem of Drèze stated above. Indeed, applying Drèze's theorem we have, in our notation, $f^{o}(B|\Pi,\Omega^{-1}) = g_1^{o}(B|\Pi,\Omega^{-1})$; but as $g_i^{o}(B|\Pi,\Omega^{-1}) = g_i^{o}(B)\propto k$, the result in 1.19 follows.

Rothenberg[1] pointed out a further difficulty of a natural conjugate approach in dealing with systems of equations. Rothenberg remarked that for the reduced form model, a natural conjugate prior density embodies unwanted restrictions on the variances of the parameters ; namely the variance-covariance matrix of the parameters of the ith equations is proportional to (is a scalar multiple of) the variance-covariance matrix of the parameters of the jth equation.

For the structural form model we can ask ourselves on the relevance of the remark of Rothenberg only for the conditional covariance matrix[2] $V(\Gamma/B)$ in view of the discussion above on the identification problem.

We can readily see that the proportionality of the variances remarked

---

[1] Rothenberg (24)

[2] It is also possible to discuss the problem with the conditional covariance matrix $V(\Gamma|B,\Sigma^{-1})$.

by Rothenberg appears also in the covariance matrix $V(\Gamma/B)$ as given in 1.21.

Indeed, the (conditional) covariance matrix of the coefficients corresponding to the exogenous variables of the ith equation is $\omega_{*ii} M^{o-1}_{zz}$, with $\omega_{*ii} = B_{i.} M^{o}_{yy.z} (B_{i.})'$, where $B_{i.}$ denotes the ith row of B. Similarly, the conditional-covariance matrix of the coefficients of the jth equation is $\omega_{*jj} M^{o-1}_{zz}$, with $\omega_{*jj} = B_{j.} M^{o}_{yy.z} (B_{j.})'$, $B_{j.}$ being the jth row of B.

Thus the two covariance matrices are equal up to a multiplication scalar . This implies the proportionality of the variances of the coefficients of the exogenous variables of the ith and j^{th} equation.

In many practical applications, the investigator will feel that this proportionality of the variances places a severe restriction on his prior information. One such case will arise for example when prior information on the equations of a model arises from independent sources (one source per equation), that are different in nature from the information contained in the sample data. A good illustration of this point is provided by the case where one combines the likelihood of a sample obtained from time-series data with prior information coming from different sets of cross-section data, each set pertaining to one particular equation in the model.

In order to overcome this severe restriction of proportionality of the variances, we may extend the parameter set given by M^o in 1.15 to a set of m^2 matrices M^o_{ij} that will reflect the fact that information for each equation is particular to the equation.[1] This is done in section 2 below.

It is interesting to note that the problem of Rothenberg can be eliminated when the model embodies prior exclusion restrictions. This is seen in section 2.3 below upon inspecting the conditional covariance matrix $V(\Gamma/B,\Sigma^{-1})$.[2]

[1] This idea is very close to the idea of generalizing the matrix of prior relative precisions when studying the reduced form model. See Tiao and Zellner (29).

[2] Unfortunately, for small values of θ, it is difficult to obtain the covariance matrix $V(\Gamma/B)$, i.e. the covariance matrix of Γ conditional on B, but unconditional with respect to Σ^{-1}.

2. The extended natural conjugate density and its properties.

2.1. The extended natural conjugate density of all the parameters of the model.

Drèze[1] has proposed the following extended natural conjugate density function :

$$(2.1.)\quad f^{\circ}(\delta,\Sigma^{-1}) \propto ||B||^{\alpha}|\Sigma^{-1}|^{\frac{\theta}{2}} \exp -\frac{1}{2} \operatorname{tr} \Sigma^{-1}\{(\Delta-\Delta^{\circ})M^{\circ}(\Delta-\Delta^{\circ})'+S^{\circ}\}$$

where M° is a positive definite symmetric matrix of dimension $m(m+n-1) \times m(m+n-1)$ having m^2 blocks, M°_{ij}, each of dimension $(m+n-1) \times (m+n-1)$, S° is an $m \times m$ positive definite symmetric matrix, Δ is defined as before[2], and Δ° is a matrix having the form

$$(2.2.)\quad \Delta^{\circ} = \begin{bmatrix} \delta_1'^{\circ} & \cdots & & 0 \\ \vdots & \delta_2'^{\circ} & & \\ \vdots & & \ddots & \\ 0 & & & \delta_m'^{\circ} \end{bmatrix}$$

An alternate notation for the expression in 2.1 is :

$$(2.3.)\quad f^{\circ}(\delta,\Sigma^{-1}) \propto ||B||^{\alpha}|\Sigma|^{-\frac{\theta}{2}} \exp -\frac{1}{2}\sum_{i,j} \sigma^{ij} \{(\delta_i-\delta_i^{\circ})'M^{\circ}_{ij}(\delta_j-\delta_j^{\circ})+s^{\circ}_{ij}\}$$

$$= ||B||^{\alpha}|\Sigma|^{-\frac{\theta}{2}} \exp -\frac{1}{2}(\delta-\delta^{\circ})'\Psi^{\circ}_{\Sigma^{-1}}(\delta-\delta^{\circ}).\exp -\frac{1}{2}\operatorname{tr} \Sigma^{-1}S^{\circ}$$

---

[1] Drèze (11).

[2] See the definitions in 1.7.

where

$$(\delta-\delta^o)' = (\delta_1'-\delta_1^{o\prime}\ \delta_2'-\delta_2^{o\prime}\ \dots\ \delta_m'-\delta_m^{o\prime})$$

and

$$\Psi^o_{\Sigma^{-1}} = \begin{bmatrix} \sigma^{11}M^o_{11} & \sigma^{12}M^o_{12} & \dots & \sigma^{1m}M^o_{1m} \\ \sigma^{21}M^o_{21} & & & \\ \vdots & & & \\ \sigma^{m1}M^o_{m1} & \dots & \dots & \sigma^{mm}M^o_{mm} \end{bmatrix}$$

If we integrate out Σ^{-1} in 2.1 we are left with the marginal density[1]:

$$f^o(\delta) \propto ||B||^{\alpha}|(\Delta-\Delta^o)M^o(\Delta-\Delta^o)'+S^o|^{-\frac{\theta+m+1}{2}}$$

(2.4.)

$$\propto ||B||^{\alpha}|(\Delta-\Delta^o)'S^{o-1}(\Delta-\Delta^o)+M^{o-1}|^{-\frac{\theta+m+1}{2}}$$

The determinant that multiplies $||B||^{\alpha}$ is the kernel of a particular conditional matricvariate-t density. Indeed suppose that each row of Δ were full, i.e. that these matrices had m(m+n-1) entries instead of (m+n-1). Then this determinant would be the kernel of an m x m(m+n-1) matricvariate-t density.[2] With the null entries, we have a density on m(m+n-1) variables conditional upon the remaining m(m-1)(m+n-1) variables being equal to their expectation.

---

[1]This can be performed if $\theta > -2$ since the density $f^o(\Sigma^{-1}/\Delta)$ is a Wishart-density with parameters $(\theta+m+1, (\Delta-\Delta^o)M^o(\Delta-\Delta^o)'+S^o)$. See Appendix , Section A.

[2]See Appendix Section B.

The expressions proposed in 2.1 and 2.4 seem well behaved. Although no explicit expression is available yet for the normalizing constant, we can insure that they are proper density functions for adequate values of θ and α, as it will be shown below.

That $f^{o}(\delta)$ is a proper density is seen from the fact that 1) It is obviously non-negative. 2) For appropriate values of θ and α, the integral $\int_{R^{m(m+n-1)}} f(\delta)d\delta$ converges. This second property will be proved by showing that under some conditions on θ, the continuous (measurable) real valued function $k(\delta)$,

$$k(\delta) \underset{def}{=} ||B||^{\alpha} |(\Delta-\Delta^{o})M^{o}(\Delta-\Delta^{o})'+S^{o}|^{-\frac{1}{2}(\theta+m+1)} , k(\delta) \geq 0,$$

is integrable in the sense that the integral both exists and is finite.

To prove that $k(\delta)$ is integrable we shall show that $k(\delta)$ is dominated by a function $\phi(\delta)$ that is integrable over $R^{m(m+n-1)}$.[1]

We shall find it convenient to distinguish the treatment with $\alpha = 0$, and with $\alpha > 0$. Theorems 2.1 and 2.2 yield sufficient conditions for the existence of the density in 2.4. in the two cases respectively.

---

[1] Thus in view of a basic property of the Riemann (and of the Lebesgue) integral, if $\phi(\delta)$ is integrable over $R^{m(m+n-1)}$, and $k(\delta) \leq \phi(\delta)$, then $k(\delta)$ is also integrable over $R^{m(m+n-1)}$. See e.g. Apostol (3) p. 431, or Kolgomorov and Fomin (21) p.67.

<u>Theorem 2.1.</u> If $\alpha = 0$, a sufficient condition for the density in 2.4. to be a proper density function is that $\theta > m+n-3$.

<u>Proof.</u> The following inequality holds for the quadratic form

$$t'[S^o+(\Delta-\Delta^o)M^o(\Delta-\Delta^o)']t :$$

$$(2.5.)\quad t'[S^o+(\Delta-\Delta^o)M^o(\Delta-\Delta^o)']\,t \geq t'S^o t + \lambda^* t'(\Delta-\Delta^o)(\Delta-\Delta^o)'t$$

where λ^* is the <u>smallest</u> eigenvalue of M^o, and where use has been made of the fact that S^o and M^o are positive definite symmetric matrices.

The inequality in 2.5. can still be rearranged as :

$$t'\{[S^o+(\Delta-\Delta^o)M^o(\Delta-\Delta^o)'] - [S^o+\lambda^*(\Delta-\Delta^o)(\Delta-\Delta^o)']\}t \geq 0$$

and the following determinantal inequality follows :[1]

$$|S^o+(\Delta-\Delta^o)M^o(\Delta-\Delta^o)'| \geq |S^o+\lambda^*(\Delta-\Delta^o)(\Delta-\Delta^o)'|$$

Accordingly :

$$(2.6.)\quad k(\delta) = |S^o+(\Delta-\Delta^o)M^o(\Delta-\Delta^o)'|^{-\frac{1}{2}(\theta+m+1)} \leq |S^o+\lambda^*(\Delta-\Delta^o)(\Delta-\Delta^o)'|^{-\frac{1}{2}(}$$

---

[1] If A and B are two positive definite symmetric matrices, and A-B is positive semi-definite symmetric, then $|A| \geq |B|$. See e.g. Dhrymes (8) p.584.

The left hand side of 2.6. is the kernel of the density in 2.4. with $\alpha = 0$.

We now show that the right hand side of 2.6. is integrable. Indeed, we first remark that

$$|S^o+\lambda^*(\Delta-\Delta^o)(\Delta-\Delta^o)'|^{-\frac{1}{2}(\theta+m+1)}$$

is proportional to :[1]

$$(2.7.)\ \int_{|R>0} |\Sigma^{-1}|^{\frac{\theta}{2}} \exp -\frac{1}{2}\operatorname{tr}\Sigma^{-1}\{S^o+\lambda^*(\Delta-\Delta^o)(\Delta-\Delta^o)'\}d\Sigma^{-1}$$

The trace in 2.7. can be rearranged as :

$$(2.8.)\ \operatorname{tr}\Sigma^{-1}\{S^o+\lambda^*(\Delta-\Delta^o)(\Delta-\Delta^o)'\} = \operatorname{tr}\Sigma^{-1}S^o+(\delta-\delta^o)'G^o(\delta-\delta^o)$$

with $$G^o = \lambda^* \begin{pmatrix} \sigma^{11}I_{m+n-1} & 0 & 0 \\ 0 & \sigma^{22}I_{m+n-1} & \\ \vdots & \vdots & \\ 0 & 0 & \sigma^{mm}I_{m+n-1} \end{pmatrix}$$

Inserting 2.8. in 2.7. we obtain

$$(2.9)\ \int_{|R>0} |\Sigma^{-1}|^{\frac{\theta}{2}} \exp -\frac{1}{2}\{(\delta-\delta^o)'G^o(\delta-\delta^o)+\operatorname{tr}\Sigma^{-1}S^o\}d\Sigma^{-1}$$

---

[1] See Appendix Section A.

Accordingly

$$(2.10.)\quad \int_R |S^o+\lambda^*(\Delta-\Delta^o)(\Delta-\Delta^o)'|^{-\frac{1}{2}(\theta+m+1)} d\delta \propto$$

$$\int_R \int_{|R>0} |G^o|^{\frac{1}{2}} |G^o|^{-\frac{1}{2}} |\Sigma^{-1}|^{\frac{\theta}{2}} \exp -\frac{1}{2}\{(\delta-\delta^o)'G^o(\delta-\delta^o)+\mathrm{tr}\ \Sigma^{-1}S^o\} d\Sigma^{-1}.$$

After integrating with respect to δ the right hand side of 2.10., we are left with

$$(2.11.)\quad \int_R |S^o+\lambda^*(\Delta-\Delta^o)(\Delta-\Delta^o)'|^{-\frac{1}{2}(\theta+m+1)} d\delta \propto \int_{|R>0} |G^o|^{-\frac{1}{2}} |\Sigma^{-1}|^{\frac{\theta}{2}} \exp -\frac{1}{2}\mathrm{tr}\Sigma^{-1}S^o d\Sigma^{-1}$$

Now $|G^o|^{-\frac{1}{2}} = \prod_{i=1}^{m} \{\sigma^{ii}\}^{-\frac{1}{2}(m+n-1)} \cdot (\lambda^*)^{\frac{-m(m+n-1)}{2}}$

$$|G^o|^{-\frac{1}{2}} |\Sigma^{-1}|^{\frac{\theta}{2}} = \left\{ \frac{|\Sigma-1|}{\prod_{i=1}^{m} \sigma^{ii}} \right\}^{\frac{1}{2}(m+n-1)} |\Sigma^{-1}|^{\frac{1}{2}(\theta-(m+n-1))} (\lambda^*)^{\frac{-m(m+n-1)}{2}},$$

and the following inequality holds

$$|G^o|^{-\frac{1}{2}} |\Sigma^{-1}|^{\frac{\theta}{2}} \le |\Sigma^{-1}|^{\frac{1}{2}(\theta-(m+n-1))} (\lambda^*)^{\frac{-m(m+n-1)}{2}}$$

since $|\Sigma^{-1}| \le \prod_{i=1}^{m} \sigma^{ii}$. [1]

Thus the integral in the right hand side of 2.11. is smaller than or equal to

---

[1] See Bellman [4] p.126.

$$(2.12)\quad (\lambda^*)^{\frac{-m(m+n-1)}{2}} \int_{|R>0} |\Sigma^{-1}|^{\frac{1}{2}(\theta-(m+n-1))} \exp - \frac{1}{2}\mathrm{tr}\ \Sigma^{-1}S^o d\Sigma^{-1}.$$

The function under the integral in 2.12. is the kernel of a Wishart density with parameters S^o and $(\theta-n+2)$. The integral in 2.12. exists provided that $\theta > m+n-3$.

The existence of the integral in 2.12. implies the existence of the integrals in 2.11., and hence the integrability of the function in the left hand side of the inequality 2.6.

It remains to be shown that $\int_R |S^o+(\Delta-\Delta^o)M^o(\Delta-\Delta^o)'|^{-\frac{1}{2}(\theta+m+1)} d\delta$ is strictly positive. But this is obvious since it is easy to find a set of positive measure overwhich $k(\delta)$ is bounded away from zero. This can be seen for example by considering $k(\delta)$ at the point δ^o. At δ^o, $k(\delta^o) > 0$, and by the continuity of $k(\delta)$, there exists a neighbourhood of δ of positive measure such that $f(\delta) > 0$.

Still with $\alpha = 0$, the mean of 2.4. is Δ^o. Indeed, using 2.4. and 2.3. we have :

$$(2.13.)\quad \begin{aligned} E[\Delta] &= E_{\Sigma^{-1}} E[\Delta/\Sigma^{-1}] \\ &= E_{\Sigma^{-1}}[\Delta^o] = \Delta^o \end{aligned}$$

The conditional expectation of Δ is the expectation of a Normal density, and it does not depend on Σ^{-1}.

Unfortunately, the unconditional variance of Δ is difficult to obtain analytically. The conditional variance is :

$$(2.14.)\quad V(\Delta/\Sigma^{-1}) = \Psi^{o-1}_{\Sigma^{-1}}$$

For θ large, and always with $\alpha = 0$, we can approximate the density in 2.4. by a Normal density with mean Δ^o and covariance matrix $\overline{\Psi}^{o-1}$, $\overline{\Psi}^o = [\overline{s}^{ij} M^o_{ij}]^{-1}$, i.e. $\overline{\Psi}^o$ is a matrix having typical block $\overline{s}^{ij} M^o_{ij}$, $\overline{s}^{ij}$ being an element of the inverse of $\overline{S} = \frac{1}{\theta+m+1} S^o$.

This can be seen in the following way.[1]

Consider the matric variate-t-density

$$(2.15.)\quad f^o(T) \propto |S^o+(T-T^o)M^o(T-T^o)'|^{-\frac{1}{2}(\theta+m+1)}$$

where T is a matrix of dimension m x m(m+n-1).

Partition T and T^o in

$$T = \begin{bmatrix} t'_{11} & t'_{12} & \cdots t'_{1m} \\ t'_{21} & & \\ \vdots & & \\ t'_{m1} & t'_{m2} & \cdots t'_{mm} \end{bmatrix} \quad \text{and } T^o = \begin{bmatrix} t^{o\prime}_{11} & t^{o\prime}_{12} & \cdots t^{o\prime}_{1m} \\ t^{o\prime}_{21} & & \\ \vdots & & \\ t^{o\prime}_{m1} & \cdots & t^{o\prime}_{mm} \end{bmatrix}$$

[1] The argumentation below follows Zellner (35) chap.8 p.21, with slight modifications.

where t_{ij} and t^o_{ij}, i,j = 1,2,...,m are (m+n-1) dimensional (column)vectors.

Thus the density in 2.4. is a conditional density of 2.15. assuming that $t_{ii} = \delta_i$, $t^o_{ii} = \delta^o_i$ i = 1,2,...,m and $t_{ik} = t^o_{ik}$ for $k \neq i$.

Now, we show that the density of the t_{ii}'s, i = 1,2,...,m conditional on $t_{ik} = t^o_{ik}$, $k \neq i$, is approximately Normal, with mean t^o_{ii}, i = 1,2,...,m and covariance matrix $\bar{\Psi}^{o\,-1}$.

If θ is large the density in 2.15. is approximately Normal and can be represented as :

$$f^o(T) \stackrel{\bullet}{\propto} \exp - \frac{1}{2} \operatorname{tr} \bar{S}^{-1}(T-T^o)M^o(T-T^o)'$$

$$\stackrel{\bullet}{\propto} \exp - \frac{1}{2}(t-t^o)'(\bar{S}^{-1} \otimes M^o)(t-t^o) \qquad (2.16.)$$

where "$\stackrel{\bullet}{\propto}$" denotes "approximately proportional" and t and t^o denote the column expansions of T' and $T^{o\prime}$ respectively.

We can rewrite 2.16. fully as

$$f^o(t) \stackrel{\bullet}{\propto} \exp - \frac{1}{2} \sum_{i,j} \bar{s}^{ij} \sum_{k,l} (t_{ik}-t^o_{ik})'M^o_{kl}(t_{jl}-t^o_{jl})$$

Therefore, we have the conditional density :

$$(2.17)\quad f^{o}(t_{11},t_{22},t_{33},\ldots,t_{mm}/t_{ik}-\overset{o}{t}_{ik}=0,\ i,k=1,2,\ldots,m,\ i\neq k)$$

$$\overset{\bullet}{\propto} \exp -\frac{1}{2}\sum_{i,j} \bar{s}^{ij}(t_{.i}-t^{o}_{.i})'M^{o}_{ij}(t_{.j}-t^{o}_{.j}).$$

The second line in 2.17. can be rearranged as a kernel of a Normal density on $t_{11},t_{22},\ldots,t_{mm}$, with means $t^{o}_{11},t^{o}_{22},\ldots,t^{o}_{mm}$, and covariance matrix $\bar{\Psi}^{o-1}$. This completes the argument.

If $\alpha > 0$, it is substantially more difficult to prove that 2.4. is a proper density function. However we are able to state the following lemma that will allow us to integrate out the parameters in Γ in 2.1. This lemma allows us also to represent the integral of the density of the parameters in B as a mixture of marginal moments of B, provided that they exist. Conditions for the existence of those marginal moments can be derived, and hence the existence of a proper density in 2.4. is proved, for the special case $M^{o}_{ij}=0$.

__Lemma 2.1.__ If (δ,Σ^{-1}) have the joint density[1] :

$$g^{o}(\delta,\Sigma^{-1}) \propto |\Sigma^{-1}|^{\frac{\theta}{2}}\exp \frac{-1}{2}\,\mathrm{tr}\,\Sigma^{-1}[(\Delta-\Delta^{o})M^{o}(\Delta-\Delta^{o})'+S^{o}],$$

with $\theta > m+n-1$, then :

[1] The definitions of δ, b_i, c_i are given in section 1.1. above. We recall that b_i contains the (m-1) parameters of the endogenous variables and c_i the n parameters of the exogenous variables, in the ith equation.

1. the density of b, $b' = (b'_1, b'_2 \dots b'_m)$, given Σ^{-1} is given by :

$$(2.18.) \quad g^o(b/\Sigma^{-1}) \propto |\Psi^{obb.c}|^{\frac{1}{2}} \exp \frac{-1}{2}(b-b^o)' \Psi^{obb.c}(b-b^o)$$

2. the density of c, $c' = (c'_1, c'_2 \dots c'_m)$, given b and Σ^{-1} is given by :

$$(2.19) \quad g^o(c|b,\Sigma^{-1}) \propto |\Psi^{occ}|^{\frac{1}{2}} \exp \frac{-1}{2}(c-\check{c}_b)' \Psi^{occ}(c-\check{c}_b)$$

and

3. the marginal density of Σ^{-1} is given by :

$$(2.20\ ) \quad g^o(\Sigma^{-1}) \propto |\Psi^o|_{\Sigma^{-1}}^{\frac{-1}{2}} |\Sigma^{-1}|^{\frac{\theta}{2}} \exp \frac{-1}{2} \operatorname{tr} \Sigma^{-1} S^o,$$

where $\Psi^{obb.c}$, Ψ^{occ} and $\check{c}$ are defined below as functions of Σ^{-1}.

If we partition M^o_{ij} $i,j = 1,2,\dots,m$, in :

$$M^o_{ij} = \begin{bmatrix} M^{obb}_{ij} & M^{obc}_{ij} \\ M^{ocb}_{ij} & M^{occ}_{ij} \end{bmatrix}$$ according to the dimensions of b_i, c_i, b_j and c_j,

we can define the matrices $\Psi^{obb} = [\sigma^{ij} M^{obb}_{ij}]$, $\Psi^{obc} = [\sigma^{ij} M^{obc}_{ij}]$, $\Psi^{ocb} = \Psi^{obc'}$ and $\Psi^{occ} = [\sigma^{ij} M^{occ}_{ij}]$, and $\Psi^{obb.c}$ is the matrix :

$$\Psi^{obb.c} = \Psi^{obb} - \Psi^{obc}(\Psi^{occ})^{-1} \Psi^{ocb} .$$

The conditional expectation $\check{c}_b$ in 2.19. is :

$$\check{c}_b = c^o - \Psi^{occ^{-1}} \Psi^{ocb}(b-b^o).$$

Proof. The term $(\delta-\delta^{o})'\Psi^{o}_{\Sigma^{-1}}(\delta-\delta^{o})$ in the first exponential of 2.3. can be written as :

$$(2.21)\quad \sum_{i,j} \sigma^{ij}(\delta_i-\delta_i^{o})'M^{o}_{ij}(\delta_j-\delta_j^{o}) =$$

$$\sum_{i,j} \sigma^{ij}[(b_i-b_i^{o})':(c_i-c_i^{o})'] \begin{bmatrix} M^{obb}_{ij} & M^{obc}_{ij} \\ M^{ocb}_{ij} & M^{occ}_{ij} \end{bmatrix} \begin{pmatrix} (b_j-b_j^{o}) \\ (c_j-c_j^{o}) \end{pmatrix} =$$

$$\sum_{i,j} \sigma^{ij}\{(b_i-b_i^{o})'M^{obb}_{ij}(b_j-b_j^{o})+(c_i-c_i^{o})'M^{ocb}_{ij}(b_j-b_j^{o})$$

$$+ (b_i-b_i^{o})'M^{obc}_{ij}(c_j-c_j^{o})+(c_i-c_i^{o})'M^{occ}_{ij}(c_j-c_j^{o})\} =$$

$$(b-b^{o})'\psi^{obb}(b-b^{o})+(c-c^{o})'\psi^{ocb}(b-b^{o})+(b-b^{o})'\psi^{obc}(c-c^{o})$$

$$+ (c-c^{o})'\psi^{occ}(c-c^{o}).$$

Completing the square and arranging terms in 2.21. yields :

$$(b-b^{o})'\psi^{obb.c}(b-b^{o})+(c-c^{o}+\psi^{occ^{-1}}\psi^{ocb}(b-b^{o}))'\psi^{occ}(c-c^{o}+\psi^{occ^{-1}}\psi^{ocb}(b-b^{o}))$$

$$= (b-b^{o})'\psi^{obb.c}(b-b^{o})+(c-\check{c}_b)'\psi^{occ}(c-\check{c}_b).$$

The determinant $|\Psi^{o}_{\Sigma^{-1}}|$ can be decomposed into $|\psi^{obb.c}||\psi^{occ}|$, by reindexing the rows and the columns of $\Psi^{o}_{\Sigma^{-1}}$ in order to have the submatrix $\psi^{obb.c}$ in the upper left corner, and ψ^{occ} in the lower right corner ; and by performing the usual decomposition of a determinant of a positive definite matrix. (Note that our reindexing involves an even number of permutations, so that the determinant does not change sign).

Thus the density $g^o(b,c,\Sigma^{-1})$ can take on the form :

$$(2.22.)\quad g^o(b,c,\Sigma^{-1}) \propto |\Psi^{o\,bb.c}|^{\frac{1}{2}}\exp - \tfrac{1}{2}(b-b^o)'\Psi^{o\,bb.c}(b-b^o)$$
$$\cdot\ |\Psi^{o\,cc}|^{\frac{1}{2}}\exp - \tfrac{1}{2}(c-\check{c})'\Psi^{o\,cc}(c-\check{c})$$
$$\cdot\ |\Psi^o_{\Sigma^{-1}}|^{-\frac{1}{2}}|\Sigma^{-1}|^{\frac{\theta}{2}}\exp - \tfrac{1}{2}\,\mathrm{tr}\ \Sigma^{-1}S^o.$$

Notice that in 2.22. the conditional density $f^o(c|b,\Sigma^{-1})$ appears as a Normal density with mean $\check{c}_b$ and variance-covariance matrix $(\Psi^{o\,cc})^{-1}$; the conditional density $f^o(b|\Sigma^{-1})$ is Normal with parameters $b^o_;(\Psi^{bb.c})^{-1}$, whereas the last line defines the kernel of an unspecified marginal density $g^o(\Sigma^{-1})$.

Using lemma 2.1. we can decompose 2.3. in the following way :

$$f^o(b,c,\Sigma^{-1}) \propto ||B||^{\alpha}.\ g^o(b/\Sigma^{-1})g^o(\Sigma^{-1})g^o(c/b,\Sigma^{-1})$$

Integrating out c we are left with :

$$(2.23.)\ f^o(b,\Sigma^{-1}) \propto ||B||^{\alpha}g^o(b|\Sigma^{-1})g^o(\Sigma^{-1})$$

Remark that 2.23. is a proper normalized density function if the integral

$$\int ||B||^{\alpha} \, g^{o}(b/\Sigma^{-1}) g^{o}(\Sigma^{-1}) \, db \, d\Sigma^{-1}$$

exists and is different from zero.

The integral above can be represented as the expectation $E[||B||^{\alpha}] = E_{\Sigma^{-1}} E_{b/\Sigma^{-1}}[||B||^{\alpha}]$ with the elements of B having the conditional density $g^{o}(b/\Sigma^{-1})$ given by 2.18. and with Σ^{-1} having the marginal density $g^{o}(\Sigma^{-1})$ given by 2.20.

<u>Theorem 2.2.</u> If M^{o} is block-diagonal, a sufficient condition for the density in 2.4. to be a proper density function is that $\theta > m + n + \alpha' - 3$ where α' is the smallest even integer greater than or equal to α.

<u>Proof.</u> We shall give a complete proof for m = 2, and then sketch the generalization for m > 2.

1. We will take as a starting point the representation

$$(2.24.) \quad ||B||^{\alpha} |(\Delta-\Delta^{o})M^{o}(\Delta-\Delta^{o})'+S^{o}|^{-\frac{1}{2}(\theta+m+1)} \propto$$

$$\int_{|R>0} ||B||^{\alpha} |\Sigma^{-1}|^{\frac{\theta}{2}} \exp -\frac{1}{2}\{(\delta-\delta^{o})'\Psi^{o}_{\Sigma^{-1}}(\delta-\delta^{o}) + tr\Sigma^{-1}S^{o}\} d\Sigma^{-1}$$

Using the deocmposition given by lemma 2.1. on the right hand side of 2.24. and integrating with respect to δ we obtain an expression proportional to $E[||B||^{\alpha}]$, where $E[||B||^{\alpha}]$ has the meaning given above.

If we replace the exponent α by α', then $||B||^{\alpha'} = |B|^{\alpha'}$, and it suffices to prove the existence of $E[|B|^{\alpha'}]$ to prove the existence of $E[||B||^{\alpha}]$, since if $E[|B|^{\alpha'}]$ exists, this implies the existence of the absolute moment $E[||B||^{\alpha}]$.[1]

2. Now $|B|^{\alpha'}$ is a polynomial expression in the elements of b ; conditionally to Σ^{-1}, b is distributed Normally with mean b^o and covariance matrix $(\psi^{o bb.c})^{-1}$. In the special case where $m = 2$, $M^o_{ij} = 0$ for $i \neq j$, the matrices $\psi^o_{\Sigma^{-1}}$ and $\psi^{o bb.c}$ take the form

$$\psi^o_{\Sigma^{-1}} = \begin{bmatrix} \sigma^{11} M^o_{11} & 0 \\ 0 & \sigma^{22} M^o_{22} \end{bmatrix}, \text{ and } \psi^{o bb.c} = \begin{bmatrix} \sigma^{11} m_1 & 0 \\ 0 & \sigma^{22} m_2 \end{bmatrix}$$

with the scalars m_i, $i = 1, 2$ equal to :

$$m_i = M^{bb}_{ii} - M^{bc}_{ii} \, {M^{cc}_{ii}}^{-1} \, M^{cb}_{ii} .$$

[1] See e.g. Feller (14) p.135.

3. In section 1.1. above, the normalization rule $\beta_{ii} = 1$, $i = 1,2$, was adopted, in that case b consists of the coefficients (β_{12},β_{21}). But, we may sometimes prefer to normalize on the first column of the matrix B ; in that case, b consists of the coefficients (β_{12},β_{22}).

Also the term $|B|^{\alpha'}$ can be represented either as :

$$(2.25.)\quad (1-\beta_{12}\beta_{21})^{\alpha'} = \sum_{i=0}^{\alpha'} (-1)^i \frac{\alpha'!}{i!(\alpha'-i)!} \beta_{12}^{\alpha'-i}\beta_{21}^{\alpha'-i} ,$$

or as :

$$(2.26.)\quad (\beta_{12}-\beta_{22})^{\alpha'} = \sum_{i=0}^{\alpha'} (-1)^i \frac{\alpha'!}{i!(\alpha'-i)!} \beta_{12}^{\alpha'-i}\beta_{22}^{i} ,$$

depending upon the normalization rule.

4. Taking conditional expectations with respect to the betas in 2.25. and 2.26. we obtain :

$$(2.27.)\quad E_{b/\Sigma^{-1}}[1-\beta_{12}\beta_{22}]^{\alpha'} = \sum_{i=0}^{\alpha'} (-1)^i \frac{\alpha'!}{i!(\alpha'-i)!} E[\beta_{12}^{\alpha'-i}/\Sigma^{-1}].E[\beta_{21}^{\alpha'-i}/\Sigma^{-$$

$$(2.28.)\quad E_{b/\Sigma^{-1}}[\beta_{12}-\beta_{22}]^{\alpha'} = \sum_{i=0}^{\alpha'} (-1)^i \frac{\alpha'!}{i!(\alpha'-i)!} E[\beta_{12}^{\alpha'-i}/\Sigma^{-1}].E[\beta_{22}^{i}/\Sigma^{-1}]$$

But the highest order moments $E[\beta_{kl}^{\alpha'}/\Sigma^{-1}]$, k, l = 1,2 are equal to[1] :

---

[1] We recall that all the even moments of a Normal-distribution can be expressed in terms of the variance. See Kendall and Stuart (20) Vol I p.60

$$E[\beta_{kl}^{\alpha'}/\sigma^{kk}] = \sum_{r=0}^{\frac{\alpha'}{2}} \frac{\alpha!(2r)!}{(\alpha'-2r)!r!2^{2r}} (\sigma^{kk}m_k)^{-r}\{\beta_{kl}^{o}\}^{\alpha'-2r}$$

Thus 2.27. can be represented as a polynomial in $(\sigma^{11}\sigma^{22})^{-1}$ of degree $\frac{\alpha'}{2}$. Therefore the unconditional expectation $E[(1-\beta_{12}\beta_{22})]^{\alpha'}$ exists as soon as $E_{\Sigma^{-1}}[\prod_{i=1}^{2}(\sigma^{ii})^{-\frac{\alpha'}{2}}]$ exists.

The expression in 2.28. is formed by a sum of monomials in $(\sigma^{11})^{-1}$ and $(\sigma^{22})^{-1}$ of degree $\frac{\alpha'}{2}$. Each monomial contains $(\sigma^{11})^{-1}$ and $(\sigma^{22})^{-1}$ with respective powers, say p and q, $p+q = \alpha'/2$. Thus we have to derive conditions for the existence of terms of the form $E_{\Sigma^{-1}}[(\sigma^{11})^{-p}(\sigma^{22})^{-q}]$ with $p+q = \frac{\alpha'}{2}$, $p, q \geq 0$.

5. Using 2.20. we can write for $E_{\Sigma^{-1}}[\prod_{i=1}^{2}(\sigma^{ii})^{-\alpha'/2}]$:

$$E_{\Sigma^{-1}}[\prod_{i=1}^{2}(\sigma^{ii})^{-\frac{\alpha'}{2}}] \propto \int_{R>0} |\Sigma^{-1}|^{\frac{\theta}{2}} \prod_{i=1}^{2}(\sigma^{ii})^{-\frac{\alpha'}{2}} |\Psi^{o}_{\Sigma^{-1}}|^{-\frac{1}{2}} \exp - \tfrac{1}{2}\mathrm{tr}\Sigma^{-1}S^{o}d\Sigma^{-1}$$

$$\propto \int_{R>0} |\Sigma^{-1}|^{\frac{\theta}{2}} \prod_{i=1}^{2}(\sigma^{ii})^{-\frac{1}{2}(\alpha'+n+1)} \exp - \tfrac{1}{2}\mathrm{tr}\,\Sigma^{-1}S^{o}d\Sigma^{-1} \qquad (2.29.)$$

where use has been made of the fact that $\Psi^{o}_{\Sigma^{-1}}$ is block diagonal.

The expression under the integral in the right hand side of 2.29. can still be rearranged as :

$$(2.30.)\quad \left\{ \frac{\Sigma^{-1}}{\prod_{i=1}^{2} \sigma^{ii}} \right\}^{-\frac{1}{2}(\alpha'+n+1)} |\Sigma^{-1}|^{\frac{1}{2}(\theta-\alpha'-n-1)} \exp - \frac{1}{2} \operatorname{tr} \Sigma^{-1} S^{o} d\Sigma^{-1}$$

Now the expression in brackets is smaller than or equal to 1, and thus the function in 2.30. is dominated by the function :

$$|\Sigma^{-1}|^{\frac{1}{2}(\theta-\alpha'-n-1)} \exp - \frac{1}{2} \operatorname{tr} \Sigma^{-1} S^{o},$$

that is the kernel of a Wishart density, if $\theta > \alpha'+n-1$.[1]

6. There remains to take the expectation with respect to Σ^{-1} of terms of the form $(\sigma^{11})^{-p}(\sigma^{22})^{-q}$, $p+q = \frac{\alpha'}{2}$, that appear in 2.28.; using 2.20., we can write

$$(2.31.)\quad E_{\Sigma^{-1}}[(\sigma^{11})^{-p}(\sigma^{22})^{-q}] \propto \int_{|R>0} (\sigma^{11})^{-p}(\sigma^{22})^{-q} \prod_{i=1}^{2} (\sigma^{ii})^{-\frac{n+1}{2}} |\Sigma^{-1}|^{\frac{\theta}{2}}$$

$$\cdot \exp - \frac{1}{2} \operatorname{tr} \Sigma^{-1} S^{o} d\Sigma^{-1} .$$

The function under the integral in 2.31. is dominated by

---

[1] See Appendix, section A for the properties of the Wishart density function.

$$(2.32.)\quad (\sigma^{11})^{-p}(\sigma^{22.1})^{-q}|\Sigma^{-1}|^{\frac{\theta-(n+1)}{2}} \exp - \frac{1}{2}\,\mathrm{tr}\,\Sigma^{-1}S^{o},$$

since $\left\{\dfrac{|\Sigma^{-1}|}{\prod_{i=1}^{2}\sigma^{ii}}\right\}^{\frac{m+1}{2}} \leq 1$, and $(\sigma^{22})^{-q} \leq (\sigma^{22.1})^{-q}$.

Let us make a change of variable in 2.32. from Σ^{-1} to Σ^{-1*} where Σ^{-1*} is defined as :

$$\Sigma^{-1*} = \begin{bmatrix} \sigma^{11} & \beta \\ \beta & \sigma^{22.1} \end{bmatrix} \quad ;\beta = \frac{\sigma^{12}}{\sigma^{11}}$$

We can decompose 2.32., using property A.2 in the Appendix, in an expression proportional to :

$$(2.33.)\quad \{(\beta+\frac{s^{o}_{12}}{s^{o}_{22}})^2 s^{o}_{22}+s^{o}_{11.2}\}^{\frac{1}{2}(\theta-n+3)}\{\sigma^{11}\}^{-\frac{2p}{2}}\{\sigma^{11}\}^{\frac{\theta-n+1}{2}}\exp-\frac{1}{2}\sigma^{11}\{(\beta+$$

$$\frac{s^{o}_{12}}{s^{o}_{22}})s^{o}_{22}+s^{o}_{11.2}\}.\{\sigma^{22.1}\}^{-\frac{2q}{2}}\{\sigma^{22.1}\}^{\frac{\theta-n-1}{2}}\exp-\frac{1}{2}\,\sigma^{22.1}s^{o}_{22}\{(\beta+$$

$$\frac{s^{o}_{12}}{s^{o}_{22}})^2\; s^{o}_{22}+s^{o}_{11.2}\}^{-\frac{1}{2}(\theta-n+3)}$$

$$= \{(\beta+\frac{s^{o}_{12}}{s^{o}_{22}})^2\; s^{o}_{22}+s^{o}_{11.2}\}^{\frac{1}{2}(\theta-n+3)}\{\sigma^{11}\}^{\frac{\theta-2p-n+1}{2}}\exp-\frac{1}{2}\,\sigma^{11}\{(\beta+$$

$$\frac{s^{o}_{12}}{s^{o}_{22}})s^{o}_{22}+s^{o}_{11.2}\}.\{\sigma^{22.1}\}^{\frac{\theta-2q-n-1}{2}}\exp-\frac{1}{2}\,\sigma^{22.1}s^{o}_{22}\{(\beta+\frac{s^{o}_{12}}{s^{o}_{22}})^2 s^{o}_{22}+$$

$$s^{o}_{11.2}\}^{\frac{1}{2}(\theta-n+3)}$$

The integral of the function in 2.33. will converge if $\theta > \max(2p + n - 1, 2q + n - 3)$. This condition is fulfilled if $\theta > \alpha' + n - 1$, since $2p + 2q = \alpha'$.

7. The results above can be generalized for the case $m > 2$. Those generalizations are shown below with the normalization rule $\beta_{ii} = 1$, $i = 1,2,\ldots,m$. The argument is similar for other normalizations.

The determinant $|B|$ can be expanded explicitely as :

$$(2.34.) \quad |B| = \sum_{P} (-1)^{f(\gamma_1,\gamma_2,\gamma_3,\ldots,\gamma_m)} \prod_{i=1}^{m} \beta_{ij_i}$$

where, we recall, $\beta_{ik} = 1$ if $i = k$, and P is the set of all permutations $(\gamma_1,\gamma_2,\ldots,\gamma_m)$ of the set of integers $(1,2,\ldots,m)$ and $f(\gamma_1,\gamma_2,\ldots,\gamma_m)$ is the number of transpositions required to change $(1,2,\ldots,p)$ into $(\gamma_1,\gamma_2,\ldots,\gamma_p)$.[1]

The determinant $|B|$ with power α' can be expanded in a polynomial expression of degree α' in the elements of b. Thus each term in the polynomial expression appears as a product of k monomials, $k \leq m$, where each monomial contains the parameters of a single equation and it is of degree lower than or equal to α'.

---

[1] The definition is taken from Anderson (1) p.335.

The expectation of any term in the polynomial can be written as proportional to :

$$(2.35.)\quad E[\prod_{i=1}^{k} S_i] = E_{\Sigma^{-1}}\, E_{b/\Sigma^{-1}}[\prod_{i=1}^{k} S_i]$$

$$= E_{\Sigma^{-1}} \prod_{i=1}^{k} E_{b/\Sigma^{-1}}[S_i]$$

where S_i, i = 1,2,...,k denotes a monomial of degree α' containing the parameters of the endogenous variables of the ith equation. The second line in 2.35. follows from the fact that conditionally on Σ^{-1}, the parameters of the ith equation are independent of those of the jth equation, $i \neq j$.

If S_i contains a single parameter, say $\beta_{ij_i}^{\alpha'}$, then 2.35. appears as :

$$(2.36.)\quad E[\prod_{i=1}^{k} S_i] = E_{\Sigma^{-1}} \prod_{i=1}^{k} E_{b/\Sigma^{-1}}[\beta_{ij_i}^{\alpha'}]$$

If k = m, we can generalize the results of step 5 above, and the condition of existence of the integral in 2.36. is that θ be greater than α'+m+n-1.

If k < m, we can use an argument simular to the argument in step 6 with either p or q = 0, and the existence of the integral in 2.36. is insured when θ greater than α'+m+n-1.

If the monomial S_i contains two or more parameters we can prove that $E[S_i]$ is bounded by a polynomial in $[\sigma^{ii}]^{-1}$ of degree $\frac{\alpha'}{2}$. We prove this for the case of two parameters. The general case is proved by repeating the same argument.

Consider a monomial S_i of the following form :

$$\beta_{ik}^{p}\ \beta_{il}^{q} \qquad p,q > 1 \quad p+q < \alpha'$$

Then taking expectations and using Schwarz' inequality we have :

$$(2.37.) \qquad E[\beta_{ik}^{p}\beta_{il}^{q}] \leq E^{\frac{1}{2}}[\beta_{ik}^{2p}]E^{\frac{1}{2}}[\beta_{il}^{2q}] =$$

$$\left\{\sum_{r=0}^{p} \frac{2p!(2r)!}{(2p-2r)!r!2^{r}} [Var(\beta_{ik})]^{r}(E(\beta_{ik}))^{2(p-r)}\right\}^{\frac{1}{2}} .$$

$$\left\{\sum_{r=0}^{q} \frac{2q!(2r)!}{(2q-2r)!r!2^{r}} [Var(\beta_{ik})]^{r}(E(\beta_{ik}))^{2(p-r)}\right\}^{\frac{1}{2}}$$

The expressions under brackets are both polynomials in $(\sigma^{ii})^{-1}$ of degree p and q respectively. The product of the two polynomials is a polynomial in $(\sigma^{ii})^{-1}$ of degree p+q. The expression on the right hand side of 2.37. can still take the form of $(\sigma^{ii})^{-\frac{p+q}{2}}$ that multiplies an expression involving <u>positive</u> powers of σ^{ii}.

Thus $\prod_{i=1}^{k} E[S_i]$ when S_i contains two or more parameters is bounded by a polynomial expression in $\prod_{i=1}^{k} [\sigma^{ii}]^{-1}$ of degree $\frac{\alpha'}{2}$. But as we have shown above the expectation exists provided that

$\theta > m+n+\alpha'-3$.

2.2. <u>The extended natural conjugate density bearing on the parameters of a model with prior exclusion restrictions.</u>

The densities in 2.1. and 2.4. can be slightly modified to treat the model with prior exact restrictions embodied in the likelihood.

Suppose that we want to put stochastic information on the μ_i unrestricted parameters of the ith equation, $i = 1,2,\ldots,m$, instead of the (m+n+1) parameters that are used when we work with a completely underidentified model.

If some elements of $(\delta_i-\delta_i^o)$ and $(\delta_j-\delta_j^o)$ are zero, the bilinear forms $(\delta_i-\delta_i^o)'M_{ij}^o(\delta_j-\delta_j^o)$ appearing in the kernels of 2.1. and 2.2. can be written more succintily in terms of the non-zero elements of $(\delta_i-\delta_i^o)$ and $(\delta_j-\delta_j^o)$ since the rows and columns of M_{ij}^o corresponding to the zero-elements drop out. In order to lighten an already, alas, heavy notation, we shall not add a new notation to distinguish the subvectors of δ_i and δ_j containing only the non-restricted elements from the vectors δ_i and δ_j.

Thus δ_i and δ_j will denote the μ_i and μ_j dimensional vectors containing the unrestricted parameters, and consequently M^o_{ij} becomes a matrix of dimension $\mu_i \times \mu_j$ with μ_i and $\mu_j \leq m+n-1$.

With this new convention the matrices $(\Delta-\Delta^o)$ and M^o have dimension $m \times \sum_{i=1}^{m} \mu_i$ and $\sum_{i=1}^{m} \mu_i \times \sum_{i=1}^{m} \mu_i$ respectively.

It is also interesting to note that the prior in 2.1. could be interpreted as the Natural-conjugate prior of a model with exact restrictions if : 1) M^o were positive semi-definite with maximum rank m+n. 2) Δ^o were a full matrix, and may not be uniquely determined. 3) S^o could possibly be a zero matrix.[1]

Nevertheless, the same type of rearrangement and of factoring as the one given in lemma 2.1. can be applied. Remark that the conditional covariance matrix $V(c/b,\Sigma^{-1})$ may not present any longer the proportionality of variances[2] since the exclusion restrictions may reindex the elements of M^o.

The main results of the preceeding section can be rephrased with minor changes, to tackle the problem of the existence of proper densities when there are prior exact restrictions.

---

[1] See also Drèze and Morales (12)

[2] I.e. the problem of Rothenberg. See section 1.2. above

We can summarize those results in the following two theorems :

Theorem 2.3. If $\alpha=0$, a sufficient condition for the density in 2.4. to be a proper density function is that $\theta > \max_i \mu_i - 2$.

Proof. Using a reasoning similar to the proof of theorem 2.1. we have the following inequality :

$$(2.38)\quad |S^o+(\Delta-\Delta^o)M^o(\Delta-\Delta^o)'|^{\frac{-1}{2}(\theta+m+1)} \leq |S^o+\lambda^*(\Delta-\Delta^o)(\Delta-\Delta^o)'|^{\frac{-1}{2}(\theta+m+1)}$$

where λ^* is the smallest root of the $\sum_i \mu_i \times \sum_i \mu_i$ matrix M^o.

The right hand side of 2.38. can be represented as proportional to

$$(2.39.)\quad \int_{R>0} |\Sigma^{-1}|^{\frac{\theta}{2}} \exp - \frac{1}{2}\{(\delta-\delta^o)'G_1^o(\delta-\delta^o)+tr\Sigma^{-1}S^o\}d\Sigma^{-1} ,$$

with δ and δ^o of dimension $\sum_i \mu_i$, and G_1^o a matrix of dimension $\sum_i \mu_i \times \sum_i \mu_i$ having the form :

$$G_1^o = \lambda^* \begin{bmatrix} \sigma^{11}I_{\mu_1} & 0 & \cdots & 0 \\ 0 & \sigma^{22}I_{\mu_2} & & \vdots \\ \vdots & & & \vdots \\ 0 & 0 & \cdots & \sigma^{mm}I_{\mu_m} \end{bmatrix}$$

The function under the integral in 2.39. can be written :

$$(2.40.)\quad |G_1^o|^{\frac{1}{2}} \exp - \tfrac{1}{2}(\delta-\delta^o)' G_1^o(\delta-\delta^o) . |\Sigma|^{-\frac{\theta}{2}} |G_1^o|^{-\frac{1}{2}} \exp - \tfrac{1}{2} tr \Sigma^{-1} S^o \propto$$

$$|G_1^o|^{\frac{1}{2}} \exp - \tfrac{1}{2}(\delta-\delta^o)' G_1^o(\delta-\delta^o) . |\Sigma^{-1}|^{\frac{\theta}{2}} \prod_{i=1}^{m} (\sigma^{ii})^{-\frac{\mu_i}{2}} \exp - \tfrac{1}{2} tr \Sigma^{-1} S^o$$

Accordingly

$$(2.41.)\quad \int_R |S^o + \lambda^*(\Delta-\Delta^o)(\Delta-\Delta^o)'|^{-\frac{1}{2}(\theta+m+1)} d\delta \propto$$

$$\int_R \int_{|R>0} |G_1^o|^{\frac{1}{2}} \exp - \tfrac{1}{2}(\delta-\delta^o)' G_1^o(\delta-\delta^o) . \prod_{i=1}^{m} (\sigma^{ii})^{-\frac{\mu_i}{2}} |\Sigma^{-1}|^{\frac{\theta}{2}} .$$

$$\exp - \tfrac{1}{2}\, tr\, \Sigma^{-1} S^o d\, \Sigma^{-1} d\delta .$$

Integrating explicitely in δ in the second line of 2.41. , we are left with

$$(2.42.)\quad \int_{|R>0} \prod_{i=1}^{m} (\sigma^{ii})^{-\frac{\mu_i}{2}} |\Sigma^{-1}|^{\frac{\theta}{2}} \exp - \tfrac{1}{2}\, tr\, \Sigma^{-1} S^o d\Sigma^{-1} .$$

The function under the integral in 2.42. can be written as

$$(2.43.)\quad \prod_{i=1}^{m} (\sigma^{ii})^{-\frac{1}{2}(\max_i \mu_i)} \prod_{i=1}^{m} \sigma^{ii\, \frac{1}{2}(\max_i \mu_i - \mu_i)} |\Sigma^{-1}|^{\frac{\theta}{2}} \exp - \tfrac{1}{2} tr \Sigma^{-1} S^o$$

$$\leq \prod_{i=1}^{m} (\sigma^{ii})^{\frac{1}{2}(\max_i \mu_i - \mu_i)} |\Sigma^{-1}|^{\frac{\theta - \max \mu_i}{2}} \exp - \tfrac{1}{2} tr \Sigma^{-1} S^o ,$$

since $\left\{ \dfrac{|\Sigma^{-1}|}{\prod_{i=1}^{m} \sigma^{ii}} \right\}^{\frac{1}{2}\max_i \mu_i} \leq 1.$

The expression $|\Sigma^{-1}|^{\frac{1}{2}(\theta-\max_i \mu_i)} \exp - \frac{1}{2} \operatorname{tr} \Sigma^{-1} S^{\circ}$ appearing in the second line of 2.43. is the kernel of a proper Wishart-density function, say $f_W(\Sigma^{-1}|S^{\circ}, \theta-\max_i \mu_i+(m+1))$ provided that $\theta > \max_i \mu_i - 2.$[1]

The integral of the left hand side in 2.43. is thus dominated by an integral proportional to the expectation

$$E\left[\prod_{i=1}^{m} (\sigma^{ii})^{\frac{1}{2}\nu_i}\right] = \int_{|R>0} \prod_{i=1}^{m} (\sigma^{ii})^{\frac{1}{2}\nu_i} f_W(\Sigma^{-1} S^{\circ}, \theta-\max_i \mu_i+(m+1))d\Sigma^{-1}$$

with $\nu_i = (\max_i \mu_i - \mu_i)$, $i = 1,2,\ldots,m$.

By repeated application of Schwarz' inequality we have that

$$(2.44.)\quad E\left[\prod_{i=1}^{m} (\sigma^{ii})^{\frac{1}{2}\nu_i}\right] \leq E^{\frac{1}{2}}\left[\sigma^{11\frac{2\nu_1}{2}}\right] E^{\frac{1}{2}}\left[\prod_{i=2}^{m} \sigma^{ii\frac{2\nu_i}{2}}\right]$$

$$\leq E^{\frac{1}{2}}\left[\sigma^{11\frac{2\nu_1}{2}}\right] E^{\frac{1}{4}}\left[\sigma^{22\frac{4\nu_i}{2}}\right] E^{\frac{1}{4}}\left[\prod_{i=3}^{m} \sigma^{ii\frac{4\nu_i}{2}}\right]$$

$$\vdots$$

$$\leq E^{\frac{1}{2}}\left[\sigma^{11\frac{2\nu_1}{2}}\right] E^{\frac{1}{4}}\left[\sigma^{22\frac{4\nu_2}{2}}\right] \ldots E^{\frac{1}{2m-1}}\left[(\sigma^{mm})^{\frac{2^{m-1}\nu_m}{2}}\right]$$

All the expectations in the rightmost hand side of the inequality in 2.44. exist, since the density of σ^{ii}, $i=1,2,\ldots,m$ is a gamma density function,[2] and moments of all orders exist for the gamma density functions.[3]

---

[1] See Appendix section A

[2] This is seen by applying property A3 in the Appendix, with r=1

[3] See Kendall and Stuart (20), Vol I p. 62

The existence of the integral in 2.44. implies the existence of the integral of the right hand side in 2.38., implying also the existence of the integral of the left hand side of 2.38.

The positivity of $\int_R |S^o+(\Delta-\Delta^o)M^o(\Delta-\Delta^o)'|^{-\frac{1}{2}(\theta+m+1)} d\delta$ is proved in the same way as in the proof of theorem 2.1.

<u>Theorem 2.4.</u> If M^o is block diagonal, a sufficient condition for the density in 2.4. to be a proper density function is that $\theta > \alpha' + \max_i \mu_i - 2$, where α' is the smallest even integer greater than or equal to α.

<u>Proof.</u> Using the same argument as in the proof of theorem 2.2., we have only to prove that :

$$E_{\Sigma^{-1}}[\prod_{i=1}^{m}(\sigma^{ii})^{-\frac{\alpha'}{2}}]$$

exists. But :

$$(2.45.)\ E_{\Sigma^{-1}}[\prod_{i=1}^{m}(\sigma^{ii})^{-\frac{\alpha'}{2}}]\propto\int_{R>0}|\Sigma^{-1}|^{\frac{\theta}{2}}\prod_{i=1}^{m}(\sigma^{ii})^{-\frac{\alpha'}{2}}|\Psi^o_{\Sigma^{-1}}|^{-\frac{1}{2}}\exp-\frac{1}{2}\mathrm{tr}\Sigma^{-1}S^o d\Sigma^{-1}$$

$$\propto\int_{R>0}|\Sigma^{-1}|^{\frac{\theta}{2}}\prod_{i=1}^{m}(\sigma^{ii})^{-\frac{\alpha'}{2}}\prod_{i=1}^{m}(\sigma^{ii})^{-\frac{\mu_i}{2}}\exp-\frac{1}{2}\mathrm{tr}\Sigma^{-1}S^o d\Sigma^{-1}$$

The non-negative continuous function under the integral in the second line of 2.43. is dominated by the function

$$(2.46.)\quad \prod_{i=1}^{m} (\sigma^{ii})^{\frac{1}{2}(\max_i \mu_i - \mu_i)} \, |\Sigma^{-1}|^{\frac{1}{2}(\theta-\alpha'-\max_i \mu_i)} \exp - \frac{1}{2} \operatorname{tr} \Sigma^{-1} S^{o} \; .$$

The function in 2.46. appears thus as the product $\prod_{i=1}^{m} (\sigma^{ii})^{\frac{1}{2}(\max_i \mu_i - \mu_i)}$ times a proper Wishart-density, provided that $\theta > \alpha' + \max_i \mu_i - 2$. The integral of 2.46. is proportional to the expectation $E[\prod_{i=1}^{m} (\sigma^{ii})^{\frac{1}{2}(\max_i \mu_i - \mu_i)}]$, which exists. The integrability of 2.46. implies the existence of 2.45.

The positivity of the integral of

$$k(\delta) = ||B||^{\alpha} |S^{o} + (\Delta-\Delta^{o}) M (\Delta-\Delta^{o})'|^{-\frac{1}{2}(\theta+m+1)}$$

is proved by the fact that there exists a set of δ of positive measure for which $k(\delta)$ is bounded away from zero since $k(\delta)$ is continuous and it is not equal to zero everywhere.

2.3. Interpretation of the extended natural conjugate density.

The analytical treatment of the densities 2.1. and 2.4. in the preceeding section is incomplete ; but we have at least derived conditions for the integrability of these densities. We have looked for conditions that insure the legitimacy of a numerical treatment, as far as the constant of integration and the first moments are concerned. In principle, there is thus no difficulty in using the prior densities in 2.1. and 2.4. ; but the following question is still open : what kind of information can this density represent adequately, or in other words, what kind of information can be summarized in the parameters of 2.1. or 2.4. If we think of the density in 2.1. as a likelihood obtained from previous samples, (according to the natural conjugate theory) we might be interested in the kind of sample that yields such a likelihood.

The task is not easy if $\alpha > 0$, since the term $||B||^{\alpha}$ introduces some awkwardness.

If $\alpha = 0$, the density in 2.1. (or the marginal density in 2.4.) reflects the fact that 1) there exist different sets of "data", each pertaining to one equation, 2) the different sets of data are related through Σ .

In classical Econometrics, the model of "seemingly unrelated regressions" of Zellner reflects those properties[1]. We may find it illuminating therefore to compare the (posterior) densities arising from the combination of a non-informative prior with the likelihood of a model of "seemingly unrelated equations"[2], with our density functions in 2.1. and 2.4.

---

[1] Zellner (34)

[2] See Tiao and Zellner (29)

To fix the ideas, we might find it tempting to investigate the form of the likelihood (and of the posterior densities) of a model consisting of m independent sets of cross-section data, each set being relevant to one of the structural equations.[1]

Les us assume that there is a well defined relationship between the specifications of the structural equations at the cross-section level and at the macro-economic level. For example, suppose that we have g observations for the ith equation, and write the model as :

$$(2.47.) \quad \tilde{y}_i = \tilde{X}_i \delta_i + \tilde{u}_i$$

The tildes indicate that $(\tilde{y}_i : \tilde{X}_i)$ are micro-economic observations, as opposed to the macro-observations $(y_i : X_i)$ in 1.5.

We suppose that there is a well-defined relationship between the parameters of the processus generating the micro-economic disturbances $\tilde{u}_{is}$, $s = 1,2,\ldots,g$, and the macroeconomic disturbances u_{it}.

We may suppose for example, that the micro-economic disturbances $\tilde{u}_{is}$, $s = 1,2,\ldots,g$, have mean zero and the variance $k_i^2 \sigma_{ii}$. Similarly, for the jth equation we would have that $\tilde{u}_{js}$, $s = 1,2,\ldots,g$, have mean zero and variance $k_j^2 \sigma_{jj}$. We assume furthermore that the covariance between $\tilde{u}_{is}$ and $\tilde{u}_{js}$ is $k_i k_j \sigma_{ij}$. Thus the matrix of

[1]This example is taken from Drèze (11)

variances and covariances $E(\tilde{\tilde{u}}\tilde{\tilde{u}}')$ can be written :

$$(2.48.)\quad E(\tilde{\tilde{u}}\tilde{\tilde{u}}') = [(K'\Sigma K) \otimes I_g]$$

$$= (K' \otimes I_g)(\Sigma \otimes I_g)(K \otimes I_g)$$

where K is the mxm diagonal matrix :

$$K = \begin{bmatrix} k_1 & 0 \cdots\cdots & 0 \\ 0 & k_2 & \vdots \\ \vdots & \vdots & \vdots \\ 0 & 0 \cdots\cdots & k_m \end{bmatrix}$$

The likelihood of the observations $\tilde{y}_i$, i=1,2,...,m is then :

$$(2.49.)\quad L(\delta,\Sigma/\tilde{y},\tilde{X}) \propto |\Sigma^{-1}|^{\frac{g}{2}} \exp - \frac{1}{2}(\tilde{y}-\tilde{X}\delta)'(K'^{-1} \otimes I)(\Sigma^{-1} \otimes I)(K^{-1} \otimes I)(\tilde{y}-$$

The likelihood can still be written, using a representation similar to 1.9. as :

$$(2.50.)\quad L(\delta,\Sigma/\tilde{y},\tilde{X}) \propto |\Sigma^{-1}|^{\frac{g}{2}} \exp - \frac{1}{2}\mathrm{tr}\ K^{-1}\Sigma^{-1}\ K'^{-1}[\tilde{Y}-\tilde{\Xi}\Delta']'[\tilde{Y}-\tilde{\Xi}\Delta']$$

The posterior density function arising from the likelihood in 2.50. combined with a non-informative prior $f(\delta,\Sigma^{-1}) \propto |\Sigma|^{\frac{m+1}{2}}$ is equal to :

$$(2.51.)\quad \tilde{f}(\delta,\Sigma^{-1}/\tilde{y},\tilde{X}) \propto |\Sigma^{-1}|^{\frac{g-(m+1)}{2}} \exp - \frac{1}{2}\ \mathrm{tr}\ \Sigma^{-1}K^{-1}{}'[\tilde{Y}-\tilde{\Xi}\Delta']'[\tilde{Y}-\tilde{\Xi}\Delta']K^{-1}$$

$$\propto |\Sigma^{-1}|^{\frac{g-(m+1)}{2}} \exp - \frac{1}{2}\ \mathrm{tr}\ \Sigma^{-1}\{[\Delta-\tilde{\Delta}]\tilde{M}[\Delta-\tilde{\Delta}]'+\tilde{S}\}$$

where

$$\tilde{M} = (K^{-1}{}' \otimes I_{m+n-1})\Xi'\Xi(K^{-1} \otimes I_{m+n-1})$$

$$\tilde{\Delta} = K^{-1}{}'Y'\Xi(K^{-1} \otimes I_{m+n-1})\tilde{M}^{-1}$$

$$\tilde{S} = K^{-1}{}'Y'YK^{-1} - \tilde{\Delta}\tilde{M}\tilde{\Delta}'$$

Integrating out Σ^{-1} in 2.51. we are left with :

$$(2.52.) \qquad \tilde{f}(\delta) \propto |(\Delta-\tilde{\Delta})\tilde{M}(\Delta-\tilde{\Delta})'+\tilde{S}|^{-\frac{1}{2}} \quad .$$

If we compare 2.52. and 2.4., we remark that the two densities are of the same type up to one difference, namely the matrix $\tilde{\Delta}$ is not a row-diagonal matrix like Δ^o.

The density in 2.52. is a "row-diagonal conditional matric variate-t-density" [1], and so is 2.4. with the particularity that Δ^o is also row-diagonal whereas in the general case it would be a "full" matrix.

Thus 2.4. can be interpreted as arising from "previous samples" generated by a seemingly unrelated model with the characteristic that the parameter Δ^o has a row-diagonal structure.

This interpretation of the density in 2.4. as the density summarizing our information coming from different sets of data in

---

[1] See Drèze and Morales (12)

an analoguous way to a seemingly unrelated model is quite satisfactory ; but there is a lack of realism embodied in the density. Indeed it is difficult to think that the disturbances at the microeconomic level are correlated in the same way as at the macroeconomic level.

A more realistic situation would be one in which we have independent information on each equation and we would want to put them in a form compatible with the density in 2.1. There is however a difficulty to incorporate independent information adequately, given that the density of Σ^{-1} conditional on δ is a Wishart-density function.[1]

For the particular case m = 2, we have however the results below. Assume that we have independent information on δ_1 and δ_2 that takes the form of a Student multivariate density function for each δ_i, i = 1,2. Assume furthermore that the exponent appearing in each density is the same for both densities. Therefore this prior information on δ_1 and δ_2 appearing under the form of a product of Student multivariate density functions would be reconciled with the

[1]We recall that the density appearing in 2.1. is $f^o(\delta,\Sigma^{-1})= f^o(\delta).f^o(\Sigma^{-1}/\delta)$ with $f^o(\Sigma^{-1}/\delta)$ a Wishart density with parameters $([(\Delta-\Delta^o)M^o(\Delta-\Delta^o)'+S^o],\theta+m+1)$.

prior in 2.4. if, and only if, M^o_{12}, s^o_{12} and α are equal to zero, so that :

$$(2.53.)\quad f^o(\delta) \propto \{(\delta_1-\delta_1^o)'M^o_{11}(\delta_1-\delta_1^o)+s^o_{11}\}^{-\frac{\theta+3}{2}}\{(\delta_2-\delta_2^o)'M^o_{22}(\delta_2-\delta_2^o)+s^o_{22}\}^{-\frac{\theta+3}{2}}$$

If we interpret 2.53. as the marginal density obtained from 2.1. by integrating out Σ^{-1}, our prior density is now fully given as :

$$(2.54.)\quad f^o(\delta,\Sigma^{-1}) \propto |\Sigma^{-1}|^{\frac{\theta}{2}} \exp -\frac{1}{2}\{\sigma^{11}[(\delta_1-\delta_1^o)'M^o_{11}(\delta_1-\delta_1^o)+s^o_{11}]$$

$$+ \sigma^{22}[(\delta_2-\delta_2^o)'M^o_{22}(\delta_2-\delta_2^o)+s^o_{22}]\}$$

Considering that our prior information bears independently on the parameters of each equation, it is convenient to perform an integrand transformation from Σ^{-1} to Σ so as to deal explicitely with σ_{11} and σ_{22}, rather than σ^{11} and σ^{22}. Furthermore, if we have no specific information about σ_{12}, it is easier to assess the implications of 2.54. in terms of $\rho^2=\sigma^2_{12}(\sigma_{11}\sigma_{22})^{-1}$, rather than σ_{12} itself.

Performing the integrand transformation from Σ^{-1} to Σ in 2.54., we can rewrite this expression as :[1]

$$(2.55.)\quad f(\delta,\sigma_{11},\sigma_{22},\sigma_{12}) \propto |\Sigma^{-1}|^{\frac{1}{2}(\theta+6)} \exp -\frac{1}{2}\left\{\frac{Q^o_{11}}{\sigma_{11}}\cdot\frac{1}{\left(1-\frac{\sigma^2_{12}}{\sigma_{11}\sigma_{22}}\right)}+\frac{Q^o_{22}}{\sigma_{22}}\cdot\frac{1}{1-\left(\frac{\sigma^2_{12}}{\sigma_{11}\sigma_{22}}\right)}\right\}$$

---

[1] See Tiao and Zellner (29), for the transformation from Σ^{-1} into Σ.

where $Q^o_{11} = ((\delta_1-\delta^o_1)'M^o_{11}(\delta_1-\delta^o_1)+s^o_{11})$

and $Q^o_{22} = ((\delta_2-\delta^o_2)'M^o_{22}(\delta_2-\delta^o_2)+s^o_{22})$.

Making a change of variable from σ_{12} to ρ^2, with Jacobian :

$$J(\sigma_{12} \curvearrowright \rho^2) = \frac{1}{2}(\rho^2)^{-\frac{1}{2}}(\sigma_{11}\ \sigma_{22})^{\frac{1}{2}} ;$$

we obtain the following representation for 2.55.

$$(2.56.)\ f^o(\delta,\sigma_{11},\sigma_{22},\rho^2) \propto (\sigma_{11}\sigma_{22})^{-\frac{1}{2}(\theta+5)}(\rho^2)^{-\frac{1}{2}}(1-\rho^2)^{-\frac{1}{2}(\theta+6)}$$

$$\cdot \exp -\frac{1}{2}\{\frac{Q^o_{11}}{\sigma_{11}} \cdot \frac{1}{(1-\rho^2)} + \frac{Q^o_{22}}{\sigma_{22}} \cdot \frac{1}{(1-\rho^2)}\} .$$

The density in 2.56. admits the following decomposition :

$$(2.57.)\ f^o(\delta_1,\delta_2,\sigma_{11},\sigma_{22},\rho^2) = f^o_1(\delta_1,\sigma_{11}/\rho^2).f^o_2(\delta_2,\sigma_{22}/\rho^2).f^o_3(\rho^2)$$

where :

$$f^o_1(\delta_1\ ,\sigma_{11}/\rho^2) \propto \sigma_{11}^{-\frac{1}{2}(\theta+5)}(1-\rho^2)^{\frac{1}{2}(\theta+3)} \exp -\frac{1}{2}\{\frac{Q^o_{11}}{\sigma_{11}(1-\rho^2)}\}$$

$$f^o_2(\delta_2,\sigma_{22}/\rho^2) \propto \sigma_{22}^{-\frac{1}{2}(\theta+5)}(1-\rho^2)^{\frac{1}{2}(\theta+3)} \exp -\frac{1}{2}\{\frac{Q^o_{22}}{\sigma_{22}(1-\rho^2)}\}$$

$$f^o_3(\rho^2) \propto [\rho^2]^{\frac{1}{2}-1}[1-\rho^2]^{\frac{1}{2}(\theta+2)-1} .$$

We may remark that (δ_1,σ_{11}) and (δ_2,σ_{22}) are independent conditionally on ρ^2 ; their densities $f_1^o(\delta_1,\sigma_{11}/\rho^2)$ and $f_2^o(\delta_2,\sigma_{22}/\rho^2)$ are Normal inverted-gamma-1 density functions ;[1] indeed :

$$(2.58.)\quad f_i^o(\delta_i,\sigma_{ii}/\rho^2)\propto[(1-\rho^2)\sigma_{ii}]^{\frac{1}{2}(n+1)} \exp - \frac{1}{2\sigma_{ii}(1-\rho^2)}(\delta_i-\delta_i^o)'M_{ii}^o(\delta_i-\delta_i^o)$$

$$.\ (1-\rho^2)^{\frac{1}{2}(\theta-n+2)}\ \sigma_{ii}^{-\frac{1}{2}(\theta-n+2)-1} \exp - \frac{s_{ii}^o}{2\sigma_{ii}(1-\rho^2)},$$

$i = 1,2.$

The marginal density $f_3^o(\rho^2) \propto [\rho^2]^{\frac{1}{2}-1}[1-\rho^2]^{\frac{1}{2}(\theta+2)-1}$ is a Beta density function[1] with parameters $p = \frac{1}{2}$ and $q = \frac{1}{2}(\theta+2)$; its two first moments are :

$$\mu_1 = \frac{1}{\theta+3}$$

$$\mu_2 = \frac{2(\theta+2)}{(\theta+3)^2(\theta+5)}$$

To specify our prior information in terms of two independent Normal-gamma densities on (δ_i,σ_{ii}) and a beta density of ρ^2 seems convenient enough. There are, however, two severe limitations : (i) the Normal-inverted gamma densities on (δ_i,σ_{ii}) are conditional upon ρ^2 ; (ii) the Beta density on ρ^2 has two parameters, of which

[1] See Raiffa and Schlaiffer (22) p.227 for the inverted-gamma-1 density, and p.216 for the Beta density function.

the first is fixed $(= \frac{1}{2})$ and the second depends upon θ ; which also appears as a parameter in the Normal-inverted gamma parts of 2.57.

These limitations carry an implication on the marginal density in 2.53. The two independent multivariate Student density functions in 2.53 have parameters $(\delta_1^o, M_{11}^o, s_{11}^o, \theta-n+2)$ and $(\delta_2^o, M_{22}^o, s_{22}^o, \theta-n+2)$ respectively, and our prior judgements on δ_1 and δ_2 can be summed up in the first two moments

$$(\delta_i^o, \frac{s_{11}^o}{\theta-n-3} M_{11}^{o\;-1}) \text{ and } (\delta_2^o, \frac{s_{22}^o}{\theta-n-3} M_{22}^{o\;-1}) \text{ respectively.}$$

Now, if we start with those moments, and reasoning backwards arrive at a density on (δ, Σ^{-1}) in the form given in 2.54., or arrive at the more convenient density on $(\delta, \sigma_{11}, \sigma_{22}, \rho^2)$ given in 2.57., we are by the same token specifying completely the distribution of ρ^2. Indeed, the unconditional variances of δ_1 and δ_2 depend on θ, and so does the complete specification of the density on ρ^2.

On the other hand, we could instead specify the distribution of ρ^2 to start with, and then specify the marginal density functions of δ_1 and δ_2, adjusting the value of s_{ii}^o to preserve the prior information on the (marginal) variances of δ_i and the specification on ρ^2. Thus s^o_{ii} will vary according to the values of θ. But, we cannot choose θ freely to express our belief on ρ^2, since θ must satisfy certain constraints if we want to have proper density functions in 2.53. In particular, with the form of the density in 2.41., and the constraints mentionned before it is not possible to specify a rectangular distribution on ρ^2.

Nevertheless, we can study the behaviour of the Beta density function for a grid of values of θ, and choose the value of θ that best reconciles the expression of our beliefs on ρ^2 with our information on δ_1, and δ_2 (as expressed in the parameters of the Student density functions). [1]

---

[1] A further difficulty is indeed the presence of θ in both Student density functions

3. Posterior distributions of the structural parameters (δ,Σ^{-1}).

3.1. The joint a posteriori density of (δ,Σ^{-1}).

In order to derive the posterior distributions, we shall use two alternate representations, each of which might have its own advantages for an adequate numerical treatment. As we shall see below, the posterior densities that we derive do not lend themselves to analytical treatment up to the point of exhibiting moments. Our aim is thus to push the analytical treatment as far as possible, by various devices, and to reduce the number of numerical integrations to be performed.

In the first representation of the posterior density function, we combine, according to Bayes' theorem, the likelihood in 1.8. and the extended natural conjugate density function as written in 2.3. ; we obtain as posterior joint density :

$$\text{(3.1.)}\quad \begin{aligned} f^{*}(\delta,\Sigma^{-1}/Y,Z) &\propto ||B||^{T+\alpha}|\Sigma^{-1}|^{\frac{T+\theta}{2}} \exp -\tfrac{1}{2}\{(\delta-\delta^{o})'\Psi^{o}_{\Sigma^{-1}}(\delta-\delta^{o})+\text{tr}\Sigma^{-1}S^{o}\} \\ &\cdot \exp -\tfrac{1}{2}(y-X\delta)'(\Sigma^{-1}\otimes I)(y-X\delta) \end{aligned}$$

The exponential part of 3.1. can be rearranged as follows :

$$\text{(3.2.)}\quad (\delta-\delta^{o})'\Psi^{o}_{\Sigma^{-1}}(\delta-\delta^{o}) + \text{tr}\Sigma^{-1}S^{o} + (y-X\delta)'(\Sigma^{-1}\otimes I)(y-X\delta)$$

$$= \delta'\Psi^o_{\Sigma^{-1}}\delta - \delta^{o\prime}\Psi^o_{\Sigma^{-1}}\delta - \delta'\Psi^o_{\Sigma^{-1}}\delta^o + y'(\Sigma^{-1}\otimes I)y - \delta'X'(\Sigma^{-1}\otimes I)y$$

$$-y'(\Sigma^{-1}\otimes I)X\delta + \delta X'(\Sigma^{-1}\otimes I)X\delta + tr\Sigma^{-1}S^o$$

$$= \delta'(\Psi^o_{\Sigma^{-1}} + \overline{\Psi}_{\Sigma^{-1}})\delta - (\delta^{o\prime}\Psi^o_{\Sigma^{-1}} + y'(\Sigma^{-1}\otimes I)X)\delta - \delta'(\Psi^o_{\Sigma^{-1}}\delta^o + X'(\Sigma^{-1}\otimes I)y'$$

$$+\delta^{o\prime}\Psi^o_{\Sigma^{-1}}\delta^o + tr\Sigma^{-1}S^o + y'(\Sigma^{-1}\otimes I)y$$

where we have used the definition of $\overline{\Psi}_{\Sigma^{-1}} = X'(\Sigma^{-1}\otimes I)X$.

Completing the square in 3.2. we find that this exponent is equal to :

$$(3.3.)\quad (\delta-\delta^*_{\Sigma^{-1}})'\Psi^*_{\Sigma^{-1}}(\delta-\delta^*_{\Sigma^{-1}}) + tr\Sigma^{-1}S^o + \delta^{o\prime}\Psi^o_{\Sigma^{-1}}\delta^o + y'(\Sigma^{-1}\otimes I)y - \delta^{*\prime}_{\Sigma^{-1}}\Psi^*_{\Sigma^{-1}}\delta^*_{\Sigma^{-1}}$$

where $\Psi^*_{\Sigma^{-1}} = (\Psi^o_{\Sigma^{-1}} + \overline{\Psi}_{\Sigma^{-1}})$,

and $\delta^*_{\Sigma^{-1}} = (\Psi^o_{\Sigma^{-1}} + \overline{\Psi}_{\Sigma^{-1}})^{-1}[\Psi^o_{\Sigma^{-1}}\delta^o + X'(\Sigma^{-1}\otimes I)y]$

The residual part $tr\Sigma^{-1}S^o + \delta^{o\prime}\Psi^o_{\Sigma^{-1}}\delta^o + y'(\Sigma^{-1}\otimes I)y$ can be rearranged in the following way :

$$(3.4.)\quad tr\Sigma^{-1}S^o + \delta^{o\prime}\Psi^o_{\Sigma^{-1}}\delta^o + y'(\Sigma^{-1}\otimes I)y - \delta^{*\prime}_{\Sigma^{-1}}\Psi^*_{\Sigma^{-1}}\delta^*_{\Sigma^{-1}}$$

$$= tr\,\Sigma^{-1}\{S^o + \Delta^o M^o \Delta^{o\prime} + Y'Y - \Delta^*_{\Sigma^{-1}}M^*\Delta^{*\prime}_{\Sigma^{-1}}\}$$

$$\underset{def}{=} tr\,\Sigma^{-1}S^*_{\Sigma^{-1}}$$

where we have defined :

$$M^{*} = M^{o} + \Xi'\Xi$$

and

$$\Delta^{*}_{\Sigma^{-1}} = \begin{bmatrix} \delta^{*\prime}_{1\Sigma^{-1}} & 0 & \cdots & 0 \\ 0 & \delta^{*\prime}_{2\Sigma^{-1}} & & \vdots \\ \vdots & & \ddots & \\ 0 & \cdots & & \delta^{*\prime}_{m\Sigma^{-1}} \end{bmatrix}$$

Remark that $M^{*} = M^{o} + \Xi'\Xi$ is a positive definite symmetric matrix, since M^{o} is positive definite and $\Xi'\Xi$ is positive semi-definite.

Substituting the expressions in 3.3. and 3.4. back in 3.1., we have the following representation for the joint posterior density of (δ, Σ^{-1}) :

$$(3.5.)\ f^{*}(\delta,\Sigma^{-1}/Y,Z) \propto ||B||^{T+\alpha}|\Sigma^{-1}|^{\frac{T+\theta}{2}}\exp -\tfrac{1}{2}(\delta-\delta^{*}_{\Sigma^{-1}})'\Psi^{*}_{\Sigma^{-1}}(\delta-\delta^{*}_{\Sigma^{-1}})+\mathrm{tr}\Sigma^{-1}S^{*}_{\Sigma^{-1}}$$

Thus the density in 3.5. appears as the product of a Jacobian term $||B||^{T+\alpha}$, times a Normal density on δ with conditional mean $\delta^{*}_{\Sigma^{-1}}$ and covariance matrix $\Psi^{*-1}_{\Sigma^{-1}}$, and a complicated distribution on Σ^{-1} that has density proportional to :

$$|\Sigma^{-1}|^{\frac{T+\theta}{2}}|\Psi^{*}_{\Sigma^{-1}}|^{-\frac{1}{2}}\exp -\tfrac{1}{2}\,\mathrm{tr}\ \Sigma^{-1}S^{*}_{\Sigma^{-1}}$$

Unfortunately, it seems quite difficult to obtain a marginal distribution of δ out of the form 3.5. Furthermore, to evaluate analytically its constant of integration or to exhibit the unconditional moments of δ may be a difficult task. But we can use the fact that conditionally on Σ^{-1} and the betas, the distribution of the gammas, i.e. of the parameters of the exogenous variables, is Normal (this can be seen by decomposing the density in 3.5. in the same way as it was done in lemma 2.1.). After integrating out the gammas, we can perform numerical integrations on the remaining parameters.

Explicitly, this can be done in the following way :

1) Decompose the Normal part of 3.5. in a (marginal) Normal density of b, and a (conditional) Normal density of c given b ; i.e. :

$$(3.6.)\quad |\Psi^{*}_{\Sigma^{-1}}|^{\frac{1}{2}} \exp - \frac{1}{2}(\delta-\delta^{*}_{\Sigma^{-1}})'\Psi^{*}_{\Sigma^{-1}}(\delta-\delta^{*}_{\Sigma^{-1}})$$

$$= |\Psi^{*bb.c}|^{\frac{1}{2}} \exp - \frac{1}{2}(b-b^{*}_{\Sigma^{-1}})'\Psi^{*bb.c}(b-b^{*}_{\Sigma^{-1}})$$

$$.\ |\Psi^{*cc}|^{\frac{1}{2}} \exp - \frac{1}{2}(c-c^{**}_{b,\Sigma^{-1}})'\Psi^{*cc}(c-c^{**}_{b,\Sigma^{-1}})$$

where $c^{**}_{b,\Sigma^{-1}} = c^{*}_{\Sigma^{-1}} - \Psi^{*cc^{-1}}\Psi^{*cb}(b-b^{*}_{\Sigma^{-1}})$, and where the rearrangement and partitioning of $\Psi^{*}_{\Sigma^{-1}}$ in Ψ^{*cc}, Ψ^{*bc}, Ψ^{*bb} are similar to those in the proof of lemma 2.1.[1]

2) Integrate out c analytically in 3.5., and integrate numerically the remaining expression, namely :

$$f^{*}(b,\Sigma^{-1}) \propto ||B||^{T+\alpha}|\Psi^{*bb.c}|^{\frac{1}{2}}\exp - \frac{1}{2}(b-b^{*}_{\Sigma^{-1}})'\Psi^{*bb.c}(b-b^{*}_{\Sigma^{-1}})$$
$$(3.7.) \qquad |\Sigma^{-1}|^{\frac{\theta+T}{2}}|\Psi^{*}_{\Sigma^{-1}}|^{-\frac{1}{2}}\exp - \frac{1}{2}\,\mathrm{tr}\,\Sigma^{-1}S^{*}_{\Sigma^{-1}}$$

For a <u>very</u> small model, and if $T+\alpha$ is not big, the <u>conditional</u> moments of b could be obtained analytically : after some heroic manipulations, they could be expressed as combinations of $b^{*}_{\Sigma^{-1}}$ and $[\Psi^{*bb.c}]^{-1}$. But, of course, to obtain the marginal moments, we have to integrate out Σ^{-1} numerically.

3) Compute the marginal moments of c by integrating $c^{**}_{b,\Sigma^{-1}}$ and $[\Psi^{*cc}]^{-1}\Psi$ in b and Σ^{-1} numerically.

These procedures apply only to small models, since we have to resort to numerical integration techniques. But it might have some advantages to integrate numerically in b and Σ^{-1} instead of integrating analytically in Σ^{-1}, and then numerically in b and c, as is done below with our alternative representation of

[1] See section 2.1. p. 39.

the posterior density function of δ and Σ^{-1}.

Combining, according to Bayes' theorem, the prior density as written in 2.1. with the likelihood as written in 1.9., yields the following posterior density

$$(3.8.)\quad f^{*}(\delta,\Sigma^{-1}/Y,Z) \propto ||B||^{T+\alpha}|\Sigma^{-1}|^{\frac{T+\theta}{2}}\exp - \frac{1}{2}\text{tr}\ \Sigma^{-1}[(\Delta-\Delta^{o})M^{o}(\Delta-\Delta^{o})'+S^{o}]$$

$$.\ \exp - \frac{1}{2}\text{tr}\Sigma^{-1}[Y-\Xi\Delta']'[Y-\Xi\Delta]$$

We can rearrange the term $Q^{*} \underset{\text{def}}{=} \{[(\Delta-\Delta^{o})M^{o}(\Delta-\Delta^{o})'+S^{o}]+[Y-\Xi\Delta']'[Y-\Xi\Delta']\}$ in the exponential part of 3.8. in the following form :

$$(3.9.)\quad Q^{*} = (\Delta-\Delta^{o})M^{o}(\Delta-\Delta^{o})'+S^{o}+(Y-\Xi\Delta')'(Y-\Xi\Delta')$$

$$= \Delta M^{o}\Delta'-\Delta^{o}M^{o}\Delta'-\Delta M^{o}\Delta^{o\prime}+\Delta^{o}M^{o}\Delta^{o\prime}+Y'Y-\Delta\Xi'Y-Y'\Xi\Delta'+\Delta\Xi'\Xi\Delta'$$

$$= \Delta(M^{o}+\Xi'\Xi)\Delta'-(\Delta^{o}M^{o}+Y'\Xi)\Delta'-\Delta(M^{o}\Delta^{o\prime}+\Xi'Y)+S^{o}+\Delta^{o}M^{o}\Delta^{o\prime}+Y'Y$$

Completing the square, we can rewrite 3.9. as :

$$(3.10.)\quad Q^{*} = (\Delta-\Delta^{*})M^{*}(\Delta-\Delta^{*})'+S^{*}$$

where $M^{*} = M^{o} + \Xi'\Xi$

$$\Delta^{*} = [\Delta^{o}M^{o}+Y'\Xi][M^{o}+\Xi'\Xi]^{-1}$$

$$S^{*} = S^{o} + \Delta^{o}M^{o}\Delta^{o\prime} + Y'Y - \Delta^{*}M^{*}\Delta^{*\prime}\ .$$

Substituting 3.10. in 3.8., we have the following representation for the posterior density function :

$$(3.11.)\quad f^{*}(\delta,\Sigma^{-1}/Y,Z) \propto ||B||^{T+\alpha}|\Sigma^{-1}|^{\frac{T+\theta}{2}} \exp -\tfrac{1}{2}\mathrm{tr}\ \Sigma^{-1}[(\Delta-\Delta^{*})M^{*}(\Delta-\Delta^{*})'+S^{*}].$$

The posterior density in 3.11. necessitates some remarks. The matrices M^{*} and S^{*} are positive definite symmetric matrices, since M^{*} is the sum of a positive definite matrix M^{o} and a positive semi-definite matrix $\Xi'\Xi$, and S^{*} can be decomposed in $U^{*\prime}\,U^{*}$ where U^{*} is of rank m. Furthermore, the matrix <u>does not</u> have any longer the same row-diagonal structure as the matrix Δ^{o}. It is a full matrix having the form :

$$\Delta^{*} = \begin{bmatrix} \delta^{*\prime}_{11} & \delta^{*\prime}_{12} & \cdots & \delta^{*\prime}_{1m} \\ \delta^{*\prime}_{21} & \delta^{*\prime}_{22} & & \delta^{*\prime}_{2m} \\ \vdots & & & \\ \delta^{*\prime}_{m1} & \delta^{*\prime}_{m2} & \cdots & \delta^{*\prime}_{mm} \end{bmatrix}$$

where a particular $\delta^{*\prime}_{ij}$ is a 1 x (m+n-1) dimensional vector defined as :

$$\delta^{*\prime}_{ij} = \sum_{k}[\delta^{o\prime}_{i}M^{o}_{ik}+y'_{i}X_{k}]M^{*kj} \quad ;$$

where M^{*kj} is a block of M^{*-1} .

If the prior density function is a proper density function, the posterior density functions in 3.5. or in 3.11. are proper density functions. This follows from the fact that the likelihood $L(\delta,\Sigma/Y,Z)$ is bounded for all δ in $R^{m(m+n-1)}$ and for all Σ in the region where Σ is positive definite symmetric. The product of a bounded likelihood function with a proper prior density function yields a proper posterior density function.

3.2. The marginal distribution of δ.

If we integrate out Σ^{-1} in 3.11 we obtain the following posterior marginal density function for δ :

$$(3.12.)\quad f^*(\delta) \propto ||B||^{T+\alpha} |(\Delta-\Delta^*)M^*(\Delta-\Delta^*)'+S^*|^{-\frac{1}{2}(T+\theta+m+1)}$$

The form of this posterior density function is similar to the form of the prior density in 2.4. The term that multiplies $||B||^{T+\alpha}$ in 3.12. presents itself as a conditional row-diagonal matric-t-density, i.e. a density on m(m+n-1) variables (the m vectors δ_{i}), conditionally upon the remaining m(m-1)(m+n-1) variables.[1]

We have thus shown with the representation in 3.12. that the prior density function in 2.4. is a closed density function, in

[1]See section 2.3. for a discussion of the conditional row-diagonal matric-t-density.

the sense that it yields a posterior density of the same form. We will state this result in the following proposition :

Proposition : The prior density in 2.4. belongs to a closed family of density functions.

Unhappily, the density in 3.12. is difficult to treat to the point of deriving its moments. The marginal distribution on the parameters of one equation for a two equations model has been derived by Drèze and Morales.[1] For a very small model, numerical integration techniques can be used. For larger models, we can, as a second best solution developped in the balance of this section, find the modal values of 3.12. through a procedure analogous to that followed for finding full information maximum likelihood estimates in the classical set-up. Some of the usual algorithms for the solution of the latter problem can be used in practical applications, as will be seen with the example shown below.

Using the definition of Q^* as given in 3.9., define

$C = \frac{1}{\theta+T+m+1} Q^*$. C can be also rewritten as :

$$C = \frac{1}{\theta+T+m+1} [\Delta M^* \Delta' - N^* \Delta' - \Delta N^{*\prime} + R^*]$$

where $N^* = \Delta^\circ M^\circ + Y'\Xi$,

[1] Drèze and Morales (12)

and $R^* = S^* + \Delta^* M^* \Delta^{*\prime}$.

A particular element of C can be written as :

$$(3.13.)\quad C_{ij} = \frac{1}{\theta+T+m+1}\,[\delta_i' M_{ij}^* \delta_j - (N_{ji}^*)'\delta_j - \delta_i' N_{ij}^* + r_{ij}^*]$$

where $N_{ij}^* = [M_{ij}^o \delta_j^o + X_i' y_j]$; $N_{ji}^* = [M_{ji}^o \delta_i^o + X_j' y_i]$

are (m+n-1) dimensional vectors and r_{ij}^* is a scalar.

Using the definitions above, the posterior density in 3.12. can be written as :

$$(3.14.)\quad f^*(\delta) \propto ||B||^{\eta_1} |C|^{-\frac{1}{2}\eta_2}$$

where $\eta_1 = T + \theta$

$\eta_2 = T + \theta + m + 1$.

In order to derive the modal values of 3.12. or 3.14., we have to solve the following system of equations

$$(3.15.)\quad f_\delta^* = \eta_1 \frac{\partial \log||B||}{\partial \delta} - \frac{\eta_2}{2}\frac{\partial \log|C|}{\partial \delta} = 0 .$$

To evaluate explicitly the vector of partial derivatives in 3.15. and the Hessian matrix below, we shall use the derivation of Rothenberg and Leenders.[1]

---

[1] Rothenberg and Leenders (23)

First, the vector of partial derivatives of one equation, $\frac{\partial \log||B||}{\partial \delta_i}$ is evaluated as $\frac{\partial \log|B|}{\partial \delta_i} = \begin{bmatrix} -b^i \\ \cdot\ \cdot \\ 0 \end{bmatrix}$ where b^i is the ith column of B'^{-1} corresponding to the non-normalized coefficients of the ith equation.[1]

Similarly, the vector of partial derivatives $\frac{\partial \log|C|}{\partial \delta_i}$ is

$$\frac{\partial \log|C|}{\partial \delta_i} = \sum_{k,l} \frac{\partial \log|C|}{\partial c_{kl}} \cdot \frac{\partial c_{kl}}{\partial \delta_i} = \sum_{k,l} c^{kl} \frac{\partial c_{kl}}{\partial \delta_i} .$$

But

$$\frac{\partial c_{kl}}{\partial \delta_i} = \begin{cases} 0 & \text{if} \quad k,l \neq i \\ 1/n_2(M^*_{il}\delta_l - N^*_{il}) & k = i,\ l \neq i \\ 1/n_2(M^*_{ik}\delta_k - N^*_{ik}) & k \neq i,\ l = i \\ 2/n_2(M^*_{ii} - N^*_{ii}) & k = i,\ l = i \end{cases}$$

and thus :

$$\frac{\partial \log|C|}{\partial \delta_i} = \frac{2}{n_2} \sum_k c^{ik}(M^*_{ik}\delta_k - N^*_{ik})$$

$$= -\frac{2}{n_2} \sum_k c^{ik}[N^*_{ik} : M^*_{ik}] \begin{bmatrix} 1 \\ \cdot\cdot \\ -\delta_k \end{bmatrix} .$$

[1] Or of the unrestricted parameters of the ith equation if additional restrictions are placed besides the normalization rule.

The vector of partial derivatives for the ith equation is :

$$\eta_1 \begin{bmatrix} -b^i \\ \cdot\ \cdot \\ 0 \end{bmatrix} + \sum_k c^{ik}[N^*_{ik} : M^*_{ik}] \begin{bmatrix} 1 \\ \cdot\ \cdot \\ -\delta_k \end{bmatrix}$$

and the full gradient vector in 3.15. can be written as :

$$(3.16.)\quad f_\delta = -\eta_1 b^+ + \begin{bmatrix} c^{11}M^+_{11} & c^{12}M^+_{12} & \cdots & c^{1m}M^+_{1m} \\ c^{21}M^+_{21} & c^{22}M^+_{22} & & \vdots \\ \vdots & & & \vdots \\ c^{m1}M^+_{m1} & \cdots & \cdots & c^{mm}M^+_{mm} \end{bmatrix} \begin{bmatrix} \delta^+_1 \\ \delta^+_2 \\ \vdots \\ \delta^+_m \end{bmatrix} = 0$$

where $b^{+\prime}$ is obtained by putting together the m vectors $(b^{i\prime}:0')$

and M^+_{ij} and δ^+_i are defined as follows :

$$M^+_{ij} = [N^*_{ij} : M^*_{ij}]$$

$$\delta^{+\prime}_i = [1 : -\delta'_i].$$

Remark that M^+_{ij} can be interpreted naturally as arising from a set of data (X^*_i, X^*_j, y^*_j) such that :

$$N^*_{ij} = X^{*\prime}_i y_j$$

$$M^*_{ij} = X^{*\prime}_i X^*_j .$$

The actual decomposition of M^*_{ij} and N^*_{ij} does not interest us but any decomposition of M^*_{ij} and N^*_{ij} in X^*_i, X^*_j, y^*_j provides an interesting analogy.

We can write M^{+}_{ij} as :

(3.17.) $M^{+}_{ij} = X^{*\prime}_{i}[y^{*}_{j}:X^{*}_{j}]$,

and the whole expression in 3.16. can be rewritten as :

$$(3.18.)\quad \eta_1 b^{+} - \begin{pmatrix} X^{*\prime}_1 & 0 & \cdots & 0 \\ 0 & X^{*\prime}_2 & & 0 \\ \vdots & & & \\ 0 & \cdots & \cdots & X^{*\prime}_m \end{pmatrix} (C^{-1}\otimes I) \begin{pmatrix} y^{*}_1 - X^{*}_1\delta_1 \\ y^{*}_2 - X^{*}_2\delta_2 \\ \vdots \\ y^{*}_m - X^{*}_m\delta_m \end{pmatrix}$$

In 3.18. we recognize a form akin to the full information maximum likelihood equations in the classical set up. Those Full Information Maximum Likelihood equations appear in a form analogous to 3.18. in, for example, Rothenberg and Leenders (23), or Rubble (26). But there are two differences :

1. In the F.I.M.L. equations $\eta_1 = \eta_2$, whereas in 3.18., $\eta_1 \neq \eta_2$. We recall that C depends upon η_2 (η_2 is embodied in C).

2. In the F.I.M.L. equations, the matrices $(y_i:X_i)$ i = 1,2,...,m are formed by deleting the columns of a common matrix X as required for identification. In other words, each matrix $(y_i:X_i)$ takes into account the unrestricted variables plus the variable subject to normalization, pertaining to the ith equation. On the other hand, in our representation 3.18., the matrices $(y^{*}_i:X^{*}_i)$, i = 1,2,...,m do <u>not</u> come from a common X^{*} matrix, in view of the

way in which the prior information comes in. In fact this difference is more apparent than real ; we can think of each matrix $(y_i^*:X_i^*)$ as being formed by m+n columns[1] of a common matrix Ξ^* of rank m(m+n). Thus, (m-1)(m+n) zero-restrictions placed on the ith equation are embodied in $(y_i^*:X_i^*)$.

To view the density in 3.11. as the density of a model with $m^2(m+n)$ variables, but with (m-1)(m+n) restrictions placed on each equation, is interesting from a classical viewpoint. Indeed, this is a simultaneous model with an identified structure, and the available algorithms to estimate the F.I.M.L. will not break down (with probability one), when they are used to obtain the modal values of the density in 3.12.

This point can become clearer by looking at the approximation to the solution of equations 3.16. The equations of 3.16. are non-linear in δ, and a close approximation to the modal value δ^{mod} is provided by :

$$(3.19.) \quad \delta^{mod} = \delta^1 - H_{\delta^1} f^*_{\delta^1}$$

where δ^1 is an arbitrary starting value, $f^*_{\delta^1}$ is the vector of partial derivatives in the r.h.s. of 3.16. evaluated at δ^1, and H is the Hessian matrix $\frac{\partial \log f^*(\delta)}{\partial\delta\partial\delta'}$ evaluated at δ^1.

[1] Or M_i+1, with $M_i < m+n-1$, if we treat a model with exact prior restrictions besides the stochastic restrictions.

To show the analogy with some of the standard algorithms used in computing the Full Information Maximum Likelihood estimator, we shall write explicitely the Hessian matrix H :

$$(3.20.)\quad H = n_1 \frac{\partial^2 \log|B|}{\partial\delta\partial\delta'} - \frac{n_2}{2} \frac{\partial^2 \log|C|}{\partial\delta\partial\delta'}$$

This matrix H consists of m^2 submatrices H_{ij} of the following form :

$$H_{ij} = n_1 \frac{\partial^2 \log|B|}{\partial\delta_i \partial\delta_j'} - \frac{n_2}{2} \frac{\partial^2 \log|C|}{\partial\delta_i \partial\delta_j'}$$

Evaluating $\frac{\partial^2 \log|B|}{\partial\delta_i \partial\delta_j'}$ yields :

$$\frac{\partial^2 \log|B|}{\partial\delta_i \partial\delta_j'} = \begin{bmatrix} -b^j b^{i\prime} & \cdots & 0 \\ \vdots & & \\ 0 & \cdots & 0 \end{bmatrix}$$

where $-b^i b^{j\prime}$ is a $(m-1)x(m-1)$ matrix having typical element $-\beta^{kj}\beta^{li}$.

The matrix $\frac{\partial^2 \log|C|}{\partial\delta_i \partial\delta_j'}$ is evaluated as :

$$(3.21.)\quad \frac{\partial^2 \log|C|}{\partial\delta_i \partial\delta_j'} = -\frac{2}{n_2} \frac{\partial}{\partial\delta_j'} \sum_k c^{ik} [N_{ik}^* : M_{ik}^*] \begin{bmatrix} 1 \\ \cdot\;\cdot \\ -\delta_k \end{bmatrix}$$

$$= \frac{2}{n_2}\{\sum_k c^{ik}\frac{\partial(-N^*_{ik}+M^*_{ik}\delta_k)}{\partial\delta'_j} + \sum_k (M^*_{ik}\delta_k - N^*_{ik})\sum_{r,s}\frac{\partial c^{ik}}{\partial c_{rs}}\frac{\partial c_{rs}}{\partial\delta'_j}\}$$

$$= \frac{2}{n_2}\{c^{ij}M^*_{ij} - \sum_k(M^*_{ik}\delta_k - N^*_{ik})\sum_{r,s}c^{ir}c^{sk}\frac{\partial c_{rs}}{\partial\delta'_j}\}$$

But

$$\frac{\partial c_{rs}}{\partial\delta'_j} = \frac{1}{n_2}\begin{cases} 0 & \text{if } r,s \neq j \\ \delta'_s M^*_{rj} - N^{*\prime}_{js} & r = j,\ s \neq j \\ \delta'_r M^*_{rj} - N^{*\prime}_{jr} & s = j,\ r \neq j \\ 2(\delta'_j M^*_{jj} - N^{*\prime}_{jj}) & r,s = j \end{cases}$$

and thus :

$$(3.22.)\quad \sum_{r,s} c^{ir}c^{sk}\frac{\partial c_{rs}}{\partial\delta'_j} = \frac{2}{n_2}\sum_r(c^{ir}c^{kj} + c^{ij}c^{kr})(\delta'_r M^*_{rj} - N^{*\prime}_{jr}).$$

Substituting 3.22. in 3.21. we obtain :

$$\frac{\partial^2 \log|C|}{\partial\delta_i\partial\delta'_j} = \frac{2}{n_2}c^{ij}M^*_{ij} - \frac{2}{n_2^2}\sum_k\sum_r(c^{ir}c^{kj} + c^{ij}c^{kr})(M^*_{ik}\delta_k - N^*_{ik})(\delta'_r M^*_{ij} - N^{*\prime}_{jr})$$

$$(3.23.)\quad = \frac{2}{n_2}c^{ij}M^*_{ij} - \frac{2}{n_2^2}[M^+_{i1}\delta^+_1 . M^+_{i2}\delta^+_2 \ldots M^+_{im}\delta^+_m]F_{ij}\begin{pmatrix}\delta^+_1 & M^+_{1j} \\ \delta^+_2 & M^+_{2j} \\ \vdots & \\ \delta^+_m & M^+_{mj}\end{pmatrix}$$

where F_{ij} is an mxm matrix with typical element

$$c^{ir}c^{kj} + c^{ij}c^{kr} .$$

Thus H_{ij} has the following form :

$$(3.24.) \quad H_{ij} = \eta_1 \begin{bmatrix} -b^j b^{i\prime} & \dots 0 \\ \vdots & \vdots \\ 0 & \dots\dots\dots 0 \end{bmatrix} - c^{ij} M^*_{ij} + \frac{1}{\eta_2} [M^+_{i1}\delta^+_1 . M^+_{i2}\delta^+_2 \dots M^+_{im}\delta^+_m] \cdot F_{ij} \begin{bmatrix} \delta^+_1 M^+_{1j} \\ \delta^+_2 M^+_{2j} \\ \vdots \\ \delta^+_m M^+_{mj} \end{bmatrix}$$

The Hessian matrix H formed by m blocks H_{ij} of the form 3.24 shows a complete analogy with the metric L used in the algorithms for F.I.M.L. estimates[1]. Slight modifications on those existing algorithms will allow us to use them to approximate the modal value of δ thourgh iterations akin to 3.19.

---

[1]See Rubble (27), Chernoff and Divinsky (6).

Appendix to Part I

Some properties of the Wishart-density function and the matric variate-t-density function

Some of the important properties of the Wishart-density function and matric variate-t-density function employed in the text are recalled here. The exposition is by no means complete and the results are stated without proof. More on those densities and proofs can be found in Anderson (1), Kaufman (19), Dickey (9), Wegge (31), Zellner (35).

A. The Wishart density function.

A (mxm) random positive definite symmetric matrix H is said to be distributed according to a Wishart distribution with parameters V and α, if H has density :

$$\text{(A.1.)}\quad f_W(H/V,\alpha) = k|H|^{\frac{1}{2}(\alpha-(m+1))} \exp -\frac{1}{2}\,\mathrm{tr}\,HV$$

$$\text{with}\quad k = |V|^{\frac{1}{2}\alpha} \frac{1}{2^{\frac{m\alpha}{2}}\,\pi^{\frac{m(m-1)}{4}} \prod_{i=1}^{m} \Gamma(\frac{1}{2}\alpha - \frac{1}{2}(i-1))},$$

$\alpha>m-1$, and V a positive definite symmetric matrix. The density in A.1. is defined by the region given by $|H| > 0$. In the text, we denote this region $\mathbb{R} > 0$.

If H and V are partitioned in

$$H = \begin{pmatrix} H_{11} & H_{12} \\ H_{21} & H_{22} \end{pmatrix}, \quad V = \begin{pmatrix} V_{11} & V_{12} \\ V_{21} & V_{22} \end{pmatrix}$$

with H_{11}, V_{11} of dimension rxr, $1 \leq r < m$, and if we define the matrix H^* as :

$$H^* = \begin{pmatrix} H_{11} & \beta \\ \beta' & H_{22.1} \end{pmatrix}$$

with $\beta = H_{11}^{-1} H_{12}$, $H_{22.1} = H_{22} - H_{21}H_{11}^{-1}H_{12}$, then the density of $f(H^*)$ can be decomposed in[1] :

$$\text{(A.2.)} \quad f(H^*) = f_W(H_{22.1}/V_{22},\alpha-r).f_W(H_{11}/F,\alpha+m-r)$$

$$.f_S(\beta/V_{11.2},V_{22},\hat{\beta},\alpha+m-r),$$

where $F = \{V_{11.2}+(\beta-\hat{\beta})V_{22}(\beta-\hat{\beta})'\}$, $\hat{\beta} = -V_{12}V_{22}^{-1}$

and $f_S(\beta/V_{11.2},V_{22},\hat{\beta},\alpha+m-r)$

stands for the matric variate-t-density on β as defined below in section B.

Thus, $H_{22.1}$ has a Wishart density with parameters $(V_{22},\alpha-r)$ and the density of H_{11} conditionally on β is Wishart with parameters $(F,\alpha+m-r)$.

---

[1] A proof can be found in Wegge (31)

In applications we are often interested in knowing the mean of the components of $\Sigma = H^{-1}$, when H has a Wishart density with parameters V, α. This mean has been evaluated to be :[1]

$$(\text{A}.3.)\quad E(\Sigma) = \frac{1}{\alpha-m-1}\,V, \qquad \alpha > m+1$$

B. The matric variate-t-density function.

A random pxq matrix is said to be matricvariate-t-distributed with parameter set (P,Q,C,m) if it has density :[2]

$$f(T) = [k(m,p,q)]^{-1}|P|^{-(m-q)/2}|Q|^{-p/2}|P^{-1}+(T-C)Q^{-1}(T-C)'|^{-\frac{m}{2}}$$

$$= [k(m,q,p)]^{-1}|Q|^{(m-p)/2}|P|^{q/2}|Q+(T-C)'P(T-C)|^{-\frac{m}{2}}$$

$$-\infty < t_{ij} < \infty$$

$$\text{where } k(m,p,q) = k(m,q,p) = \pi^{pq/2}\,\frac{\prod_{i=1}^{q}\Gamma(\frac{m-p-i+1}{2})}{\prod_{i=1}^{q}\Gamma(\frac{m-i+1}{2})},$$

$m > p+q-1$ and P , Q are positive definite symmetric.

[1]See e.g. Kaufman (19)

[2]See Dickey (9) , Zellner (35)

If we partition

$$T = \begin{pmatrix} T_1 & T_2 \\ p \times q_1 & p \times (q-q_1) \end{pmatrix}$$

$$T' = \begin{pmatrix} X_1 & X_2 \\ q \times p_1 & q \times (p-p_1) \end{pmatrix}$$

$$P = \begin{bmatrix} P_{11} & P_{12} \\ p_1 \times p_1 & p_1 \times (p-p_1) \\ P_{21} & P_{22} \\ (p-p_1) x p_1 & (p-p_1) x (p-p_1) \end{bmatrix}, \quad Q = \begin{bmatrix} Q_{11} & Q_{12} \\ q_1 \times q_1 & q_1 \times (q-q_1) \\ Q_{21} & Q_{22} \\ (q-q_1) x q_1 & (q-q_1) x (q-q_1) \end{bmatrix},$$

and suppose $C = 0$, then :

i. The conditional distribution of X_1 given X_2, is matric variate-t with parameters P_{11}, $Q+X_2' P_{22.1} X_2$, m.

ii. The conditional distribution of T_1 given T_2 is matric variate-t with parameters $(P^{-1}+T_2 Q_{22}^{-1} T_2')^{-1}$ $Q_{11.2}$, $T_2 Q_{22}^{-1} Q_{21}$, m.

iii. The marginal distribution of X_2 is matric variate-t with parameters $P_{22.1}$, Q, 0, $m-p_1$.

iv. The marginal distribution of T_2 is matric variate-t with parameters P, Q_{22}, 0, $m-q_1$.

If $m > p+q+1$ in B.1, then[1]

(B.2.) $E[T] = C \qquad E[T'] = C'$

(B.3.) $V[T] = \frac{1}{m-p-q-1} (P^{-1} \otimes Q)$

---

[1]See Kaufman (19)

Part II

Empirical Illustration of a Bayesian Full Information Analysis. The Analysis of the Belgian Beef Market.

1. The model and the a priori information.

1. 1. The model of Calicis.

For a numerical application of the developments of Part I, we have considered a modified version of a model of the retail markets of beef and pork in Belgium proposed by Calicis[1].

The original model of Calicis runs as follows :

$$(1.1)\quad \begin{aligned} x_1 &= \delta_{13}p_1 + \delta_{15}n_1 + \delta_{10} + u_1 \\ x_1 &= \delta_{23}p_1 + \delta_{24}p_2 + \delta_{27}y + \delta_{20} + u_2 \\ x_2 &= \delta_{34}p_2 + \delta_{36}n_2 + \delta_{30} + u_3 \\ x_2 &= \delta_{24}p_1 + \delta_{44}p_2 + \delta_{47}y + \delta_{40} + u_4 \end{aligned}$$

where 1 refers to beef, 2 refers to pork and x = quantity consumed per capita, p = price, n = cattle stock (number of heads measured at the beginning of each period), per capita and y = national income, per capita. The unobservable disturbances u_1, u_2, u_3, u_4 have zero mean and covariance matrix Σ.

The variables are in log form. The price and income variables have been deflated by an index of consumer prices. The exogenous variables are n_1, n_2, y.

[1] Calicis (5) p. 103

The parameters of the equations in the model 1.1 have been estimated by Calicis, using Belgian annual data for 1950-1965, (i.e 16 observations) with a two stage procedure that yields: [1]

$$
(1.2)\qquad
\begin{aligned}
\hat{x}_1 &= \underset{(0.183)}{0.419}\, p_1 + \underset{(0.145)}{1.412}\, n_1 + \underset{(0.892)}{3.214} \\
\hat{x}_1 &= -\underset{(3.454)}{1.260}\, p_1 - \underset{(17.552)}{0.481}\, p_2 + \underset{(1.680)}{1.905}\, y + \underset{(38.775)}{6.270} \\
\hat{x}_2 &= \underset{(2.228)}{2.5804}\, p_2 - \underset{(0.4040)}{0.0876}\, n_2 - \underset{(9.0253)}{6.6869} \\
\hat{x}_2 &= \underset{(0.503)}{0.294}\, p_1 - \underset{(0.255)}{0.614}\, p_2 + \underset{(0.245)}{0.313}\, y + \underset{(5.649)}{1.041}
\end{aligned}
$$

The values obtained in 1.2 call for the following remarks. The supply function for beef appears well behaved. The standard deviations are small relative to the estimates of the parameters. The high elasticity with respect to n_1 (the number of bovins) is somewhat puzzling. It can be explained, tentatively, by the fact that Belgian farmers, following government and Common Market guidelines have increased meat production and reduced milk production.

---

[1] The estimates given in 1.2 differ somewhat from those given by Calicis. We have proceeded to a reestimation in order to obtain the standard deviations of the estimates since they were not given in the paper of Calicis due to programming difficulties at the time of the publication. The values of the estimated elasticities in the first, second and fourth equations are identical. There is a printing error in Calicis'paper for the cross-elasticity in the second equation. Calicis'third equation seems to involve a computational error. We are grateful to Mrs Vuylsteke of the Louvain Computing Center for making her two-stage least squares program available to us.

In their study of the supply and demand for agricultural products in Belgium, Desaeyere et al.[1] explain that in the scope of European Economic Community agreements, the market possibilities and specially the foreign market possibilities for dairy produce are restricted. On the other hand, domestic demand for milk and dairy produce has been rather constant since 1959. The cattle stock increases will have thus, primarily an effect on the meat supply.

The estimates of the elasticities in the demand function seem reasonable, but their standard deviations are very large. The direct price and income elasticities are rather high, but they are not unreasonable. High price and income elasticities for meats are frequently found in the litterature.[2]

The cross elasticity with respect to the price of pork does not have the right sign, but its large standard deviation makes it difficult to attach to it any explanatory value <u>within</u> the context of the model estimated in 1.2.

As we are interested only in the beef market we shall neglect the remaining equations.

---

[1] Desayere, Stuyck, Van Broekhoven and Van Haeperen (7). p. 193

[2] For comments on price and income elasticities for meats see e.g Desayere et al. (7) p. 86,87 and 101, W.G. Tomek and W.W. Cochrane (30) p. 725, Faure (13) as quoted by Calicis (5) p. 106

The model of Calicis is , according to the best of our knowledge, the only simultaneous model that has been used to study the Belgian beef market. Furthermore, the results are sensible, and alternate specifications and methods of estimation can be judged in this frame of reference.

1.2. Two equations models for the Belgian beef market.

Unhappily, for a feasible numerical treatment of our Bayesian analysis the complete model 1.1 is too large. We are led to the analysis of a smaller model since we have to resort to numerical integration in the posterior analysis. We propose instead the following simpler models of the retail beef market :

$$(1.3) \qquad \begin{aligned} x &= \beta_{12}p + \gamma_{11}n + \gamma_{10} + u_1 \\ x &= \beta_{22}p + \gamma_{22}y + \gamma_{20} + u_2 \end{aligned}$$

The variables are defined as above. Remark that we have dropped out the variables pertaining to pork meat. Two alternate specifications can still be employed in analysing the model 1.3. The first specification retains the variables in 1.3 in log form, whereas the second specification considers the model in 1.3 as simply linear in the variables.

The exogenous variables in 1.3 are n and y. According to the classical classification the system in 1.3 is just identified, and a consistent estimation will be provided by indirect least-squares[1].

The estimation of 1.3 with the variables in log form yields :

(1.4)

$$\hat{x} = \underset{(0.1861)}{0.4467}\ p + \underset{(0.1466)}{1.40}\ n + \underset{(0.9017)}{3.0827}$$

$$\hat{x} = \underset{(0.3227)}{-1.9738}\ p + \underset{(0.1267)}{1.4676}\ y - \underset{(0.5372)}{4.3669}$$

whereas the estimation of 1.3 with the variables in natural form yields :

(1.5)

$$\hat{x} = \underset{(0.0544)}{0.1435}\ p + \underset{(11.3083)}{111.5518}\ n - \underset{(3.0938)}{16.9143}$$

$$\hat{x} = \underset{(0.1223)}{-0.6876}\ p + \underset{(0.000073)}{0.000725}\ y + \underset{(5.1420)}{34.5628}$$

The elasticities and their standard deviations computed at the mean values are :

(1.6) $$\eta^{s}_{xp} = \underset{(0.16)}{0.43} \qquad \eta_{xn} = \underset{(0.14)}{1.35}$$

in the supply equation, and :

(1.7) $$\eta^{d}_{xp} = \underset{(0.37)}{-2.08} \qquad \eta_{xy} = \underset{(0.15)}{1.49}$$

---

[1]We recall that indirect least-squares estimates coincide with two-stage least squares estimates in case of just-identification , Goldberger (16) p. 334 . The parameters in 1.4 and 1.5 have been computed with a two-stage least squares computer program .

in the demand equation .

The estimates in 1.4 and 1.5 (and the elasticity estimates in 1.6 and 1.7) are rather surprising . Their asymptotic standard deviations are very small when compared with the values of the estimates and especially with the values of the standard deviations in 1.2. The latter remark is especially true for the demand equation.

The surprisingly high precision of the estimates in 1.4 and 1.5 deserves further investigation. We hope that some light will result from our analysis of the sensitivity of the posterior density function to several different prior specifications.

The elasticity estimates (and their estimated standard deviations) in 1.4 are very similar to the elasticity estimates (and their standard deviations) given in 1.6 and 1.7 in both the supply and the demand equation.

Moreover, the estimated elasticities of the supply equation in 1.6 and 1.7 are not very different from the elasticities estimated by Calicis. On the other hand the price elasticities in 1.4 and 1.7 have a higher value, and in fact, they have a higher value than we may want to accept. As for income, our elasticity in the neigborhood of 1.50, may appear more acceptable than the elasticity of 1.91 estimated by Calicis.

For our Bayesian analysis, a treatment of model 1.3 seems more adequate. We will base our prior information on the analysis of microeconomic data, and the correspondence between the microeconomic parameters and the macroeconomic parameters is much more straightforward when using natural numbers than when using logs. In using a model with variables in log form, there are problems of interpretation of the prior density function on the parameters arising, essentially, from aggregation problems.

Remark that if elasticities are used as the relevant parameters in interpreting the results, of for decision-making, the models in 1.4 and 1.5 behave equally well. The estimated constant elasticities (and their estimated standard deviations) of 1.4 are very similar to the estimated elasticities (and their estimated standard deviations) computed at the sample means on the basis of model 1.5.

1.3. <u>The likelihood function and the prior density function.</u>

If we assume that the disturbances of model 1.3 are distributed normally with mean-zero and variance covariance matrix $(\Sigma \otimes I)$, the likelihood function can take on the form[1]:

$$(1.8) \qquad L \propto ||B||^{T} |\Sigma^{-1}|^{T/2} \exp -\frac{1}{2} \operatorname{tr} \Sigma^{-1} AMA'$$

---

[1] Remark that we normalize on the first column of B, whereas in Part I we dealt usually with the normalization rule $\beta_{ii} = 1$, i = 1, 2, ..., m. However, all the developments of Part I carry over with minor modifications.

With :

$$A = \begin{bmatrix} 1 & -\beta_{12} & -\gamma_{11} & 0 & -\gamma_{10} \\ 1 & -\beta_{22} & 0 & -\gamma_{22} & -\gamma_{20} \end{bmatrix}$$

$$B = \begin{bmatrix} 1 & -\beta_{12} \\ 1 & -\beta_{22} \end{bmatrix}$$

$$M = (x\ p\ n\ y\ \iota)'(x\ p\ n\ y\ \iota)$$

and where x, p, n, y and ι are T-dimensional vectors, ι being a vector composed entirely of ones.

Our prior information should bear on the parameters β_{12}, β_{22}, γ_{11}, γ_{22}, γ_{10} and γ_{20}, but information on the constant term parameters γ_{10} and γ_{20} is very difficult to obtain and very imprecise. We wish thus to remain entirely non-informative on them. Furthermore, for convenience in the computation of the posterior moments, we might wish to get rid of them. This is done in the following way :

1. Introduce the following decomposition of the prior density function :

$$(1.9) \qquad f^{\circ}(\beta_{12},\beta_{22},\gamma_{11},\gamma_{22},\gamma_{10},\gamma_{20},\Sigma^{-1})$$

$$= f^{\circ}(\gamma_{10},\gamma_{20}).f^{\circ}(\beta_{12},\beta_{22},\gamma_{11},\gamma_{22},\Sigma^{-1})$$

We assume that $f^{\circ}(\gamma_{10},\gamma_{20}) \propto k$, to express our lack of information.

2. Partition the matrices A and M in the following way :

$$A = (A_1 : A_2) \qquad M = \begin{bmatrix} M_{11} & M_{12} \\ M_{21} & M_{22} \end{bmatrix}$$

with :

$$A_2 = \begin{bmatrix} \gamma_{10} \\ \gamma_{20} \end{bmatrix} \qquad M_{12} = M_{21}' = \begin{bmatrix} x'\iota \\ p'\iota \\ n'\iota \\ y'y \end{bmatrix}$$

$$M_{22} = \iota'\iota = T$$

and rewrite 1.8 as :

$$(1.10) \qquad L \propto ||B||^T|\Sigma^{-1}|^{T/2}\exp-\frac{1}{2}\mathrm{tr}\Sigma^{-1}[A_1M_{11.2}A_1' + (A_2 + A_1M_{12}M_{22}^{-1})M_{22}(A_2 + A_1M_{12}M_{22}^{-1})']$$

with : $M_{11.2} = M_{11} - M_{12}M_{22}^{-1}M_{21}$

3. Combining 1.9 with 1.10 we obtain the following posterior density function :

$$(1.11) \qquad f^*(\beta_{12},\beta_{22},\gamma_{11},\gamma_{22},\gamma_{10},\gamma_{20},\Sigma^{-1}) \propto f^o(\beta_{12},\beta_{22},\gamma_{11},\gamma_{22},\Sigma^{-1})||B||^T|\Sigma^{-1}|^{\frac{T-1}{2}}\exp-\frac{1}{2}\mathrm{tr}\{\Sigma^{-1}A_1M_{11.2}A_1'\}.|\Sigma^{-1}|\exp-\frac{1}{2}\mathrm{tr}\Sigma^{-1}[A_2 + A_1M_{12}M_{22}^{-1}]M_{22}[A_2 + A_1M_{12}M_{22}^{-1}]'$$

Integrating out γ_{10} and γ_{20} in 1.11, we obtain the following posterior[1] :

$$(1.12) \qquad f^{*}(\beta_{12},\beta_{22},\gamma_{11},\gamma_{22}) \propto f^{o}(\beta_{12},\beta_{22},\gamma_{11},\gamma_{22},\Sigma^{-1})$$
$$||B||^{T}|\Sigma^{-1}|^{\frac{T-1}{2}} \exp-\tfrac{1}{2}\mathrm{tr}\Sigma^{-1}A_1M_{11.2}A_1'$$

Remark that $M_{11.2}$ is the matrix of sums of cross-product deviations from the sample means.

It is in the form 1.12 that we shall do the prior and posterior analysis. In other words, our prior density function will be defined on the parameters of the "concentrated" model[2] with likelihood proportional to :

$$(1.13) \qquad ||B||^{T}|\Sigma^{-1}|^{\frac{T-1}{2}} \exp-\tfrac{1}{2}\mathrm{tr}\Sigma^{-1}A_1M_{11.2}A_1'$$

Using the developments of Part I, our prior and posterior analysis will deal mainly with prior densities of the following type :

$$(1.14) \qquad f^{o}(\beta_{12},\beta_{22},\gamma_{11},\gamma_{22},\Sigma^{-1}) \propto ||B||^{\alpha}|\Sigma^{-1}|^{\frac{\theta}{2}}$$
$$\exp-\tfrac{1}{2}\mathrm{tr}\Sigma^{-1}[(\Delta-\Delta^{o})M^{o}(\Delta-\Delta^{o})' + S^{o}]$$

---

[1] Notice that the second line in 1.11 is the kernel of a Normal density on A_2 with mean $-A_1M_{12}M_{22}^{-1}$ and covariance matrix $(\Sigma\otimes M_{22}^{-1})$.

[2] "Concentrated" with respect to the constant terms, or model with variables in deviations to the means.

where $(\Delta-\Delta^{o})$ takes the form :

$$\begin{bmatrix} \beta_{12}-\beta_{12}^{o} & \gamma_{11}-\gamma_{11}^{o} & 0 & 0 \\ 0 & 0 & \beta_{22}-\beta_{22}^{o} & \gamma_{22}-\gamma_{22}^{o} \end{bmatrix}$$

The posterior densities arising from the prior densities of the type 1.14 will be compared with posterior densities arising from alternate priors.

The likelihood function, as in the second line of 1.12, and using Calicis' data, takes on the form :

$$(1.15) \qquad L \propto ||B||^{16}|\Sigma^{-1}|^{\frac{15}{2}} \exp-\frac{1}{2}\mathrm{tr}\Sigma^{-1} \begin{matrix} 1 & -\beta_{12} & -\gamma_{11} & 0 \\ 1 & -\beta_{22} & 0 & -\gamma_{22} \end{matrix}$$

$$\begin{bmatrix} .916056\times 10^{2} & .912972\times 10^{2} & .652853 & .212508\times 10^{6} \\ .912972\times 10^{2} & .243279\times 10^{3} & .594991 & .342839\times 10^{6} \\ .652853 & .594991 & .508686\times 10^{-2} & .146374\times 10^{4} \\ .212508\times 10^{6} & .342839\times 10^{6} & .146374\times 10^{4} & .617837\times 10^{9} \end{bmatrix} \begin{bmatrix} 1 & 1 \\ -\beta_{12} & -\beta_{22} \\ -\gamma_{11} & 0 \\ 0 & -\gamma_{22} \end{bmatrix}$$

For ulterior developments we may conveniently write 1.15 as[1] :

$$L \propto ||B||^{16}|\Sigma^{-1}|^{\frac{15}{2}} \exp-\frac{1}{2}\,\mathrm{tr}\Sigma^{-1}[Y'Y + \Delta\Xi'\Xi\Delta' - \Delta\Xi'Y - Y'\Xi\Delta']$$

[1] See section 1.2 of Part I

with :

$$\Xi'\Xi = \begin{bmatrix} .243279 \times 10^3 & .594991 & .243279 \times 10^3 & .342839 \times 10^6 \\ .594999 & .508686 \times 10^{-2} & .544991 & .146374 \times 10^4 \\ .243279 \times 10^3 & .594991 & .243279 \times 10^3 & .342839 \times 10^6 \\ .342839 \times 10^6 & .146374 \times 10^4 & .342839 \times 10^6 & .617837 \times 10^9 \end{bmatrix}$$

$$\Xi'Y = \begin{bmatrix} .912972 \times 10^2 & .912972 \times 10^2 \\ .652852 & .652852 \\ .912972 \times 10^2 & .912972 \times 10^2 \\ .212508 \times 10^6 & .212508 \times 10^6 \end{bmatrix}$$

$$Y'Y = \begin{bmatrix} 91.605637 & 91.605637 \\ 91.605637 & 91.605637 \end{bmatrix}$$

$$\Delta = \begin{bmatrix} \beta_{12} & \gamma_{11} & 0 & 0 \\ 0 & 0 & \beta_{22} & \gamma_{22} \end{bmatrix}$$

1.4. A description of the sources of prior-information

The prior information on the parameters of the model 1.15 arises from independent sets of data, the first set pertaining to the demand equation and the second to the supply equation. The characteristics of those informations are described below.

- To specify the prior density function on the parameters of the supply equation we need information on the coefficients of the price and cattle stock. Information on the price coefficient of a supply equation is hard to obtain from cross-section data. In view of our illustrative purposes, we shall remain essentially non-informative about this coefficient. This will be done by attributing a large variance to this price coefficient. This procedure will also simplify our analysis of the role of prior information (since a single coefficient in the supply equation is involved).

For the cattle stock coefficient, we have obtained information from cross-section data.

The Institute of Agricultural Economics of the Belgian Ministry of Agriculture, hereafter called I.E.A., keeps detailed accounts and relevant economic information on a number of farms distributed over all the agricultural regions of Belgium[1]. Those records exist since 1964. The number of farms with accounts being recorded by the I.E.A. has increased from year to year. For our analysis, we have used the data covering the period May 1967-May 1968[2].

---

[1]Belgium is divided in 13 agricultural regions which are suited for different agricultural activities. The partitioning in those regions has been established by public servants taking into account climatological, geological and geographic characteristics. See (17)

[2]The choice of this period was motivated by the quality of the data. Not only more farms were available in the sample, but there was more information for each farm than in the preceeding periods.

To judge the quality of the information furnished by these data, the following remarks are necessary :

1. The I.E.A. had selected 913 farms for its sample. These farms in the sample have not been chosen at random, but on a "good management" criterion. All have 5ha ($\sim$ 12.5 acres) or more of surface.[1]

2. The I.E.A. collects data on the number of heads of cattle owned by the farms at the beginning and at the end of the period in consideration, and on the number of cows, oxen, bulls, heifers and calves sold by the farmers. In order to have a better approximation of the number of cattle heads actually sold to the slaughterhouses, we discarded the farms of the regions where production is mainly of the "lean cattle" type[2]. We considered only the farms in the Dunes and Polders, Lime, Sand-Lime, Famenne and Condroz regions.

3. The records do not give directly the quantities of meat supplied. A quantity of meat variable has been constructed by multiplying the number of heads by an average yield in meat. The weights are 277.65 Kg.

---

[1] According to the General Census of the Agriculture that took place in 1959, 40.17% of the professional agricultural farms in Belgium had a surface of less than 5ha. The percentage has most likely diminished since 1959. See Annuaire Statistique de Belgique (18) and Desaeyere et al (7) p.30.

[2] That type of cattle is sold to farmers in regions where the fattening process takes place.

for a cow and 281.491 Kg. for other types of cattle.[1]

4. The data can yield information on the influence of the cattle stock on the quantity of meat supplied nationally and at the wholesale level ; instead, we would like to have information on the influence of the cattle stock on the quantities supplied at the retail level. Fortunately, most of the demand at retail is satisfied by domestic production, and furthermore the exported fraction of the production is low. The imports and exports of meat depend largely on accidental factors like weather conditions, and conditions in foreign countries.

From the sample of 913 farms, we have made a further selection of 45 farms. The 45 farms in our sample have been chosen randomly among the farms in the agricultural regions mentioned in point 2. An equation of the form $x = \alpha + \beta n$, where x is the quantity of meat, and n the cattle stock present in the farms at the beginning of the period, was fitted to the data by least squares.

---

[1]The weights have been obtained from data from the Annuaire Statistique de Belgique (18) 1968, p.226. This procedure is indeed very rough. Remark nevertheless, that quantities in the time series furnished by the National Institute of Statistics are obtained in a not very different fashion. This is due to the fact that in many Belgian slaughterhouses, the weights of the animals are no longer registered. The quantities of meat are obtained by multiplying the numbers of animals by an average weight and the yield in meat is calculated finally, by applying a fixed technical coefficient to the product number of animals - average weight.

This yielded :

$$(1.16) \qquad x = \underset{(523.077)}{101.834} + \underset{(12.459)}{109.687}n \qquad R^2 = 0.64$$

The elasticity with respect to the cattle stock (number of heads) computed at the sample means is :

$$(1.17) \qquad \eta_{xn} = \underset{(0.11)}{0.97}$$

The value of $\frac{\partial x}{\partial n}$ in 1.16 is surprisingly close to the value of the corresponding coefficient in 1.5, as estimated from time-series data. The elasticities differ, however, as between the two sets of data (time-series or cross-section). An explanation is found in the fact that the ratio $\overline{x}/\overline{n}$ computed with cross-section data is larger than the same ratio computed with time-series data. This is possibly due to the fact that in the time-series the values of x and n reflect the production and the stock of all regions, regardless of their degree of specialization in the production of cattle for meat. On the other hand, the cross-section data come only from the regions specialized in meat production.[1]

Information on the price coefficient of the demand equation in 1.3

---

[1]According to the General Census of Agriculture of 1959, 16.23% of the farms with cattle production were specialized in the production of cattle for meat in the Dunes and Polders, Lime, Sand-Lime, Famenne and Condroz regions compared to 11.53% for the country as a whole.

is, again, difficult to obtain from cross-section data. There is however abundant time-series literature on the retail price elasticity of demand[1], but not on the marginal value $\frac{\partial x}{\partial p}$. Furthermore those studies are mostly single equation models with rather different specifications from our specification in 1.3, and it is difficult to attach the same meaning to the price coefficient in 1.3 and to the price coefficients in those models. Moreover, our posterior analysis will be greatly facilitated if we remain non-informative on that price coefficient.

Information on the income coefficient comes from cross-section data. The Belgian National Institute of Statistics (hereafter called N.I.S.) undertook a budget enquiry from December 1960 to January 1961 of a sample of 1587 households[2]. The enquiry recorded detailed information on quantities and expenditures for consumption goods as well as on income for each of the households. The information was recorded on a fortnightly basis, with a total of 29 fortnights.

In order to compute our estimate of the income coefficient we have used a smaller sample.

---

[1] See e.g. Tomek and Cochrane (30). It is interesting to note that Tomek and Cochrane, using U.S. quarterly data from 1948-58, find a long run price elasticity in the range -.89, -1.0 .

[2] The 1587 households were selected using cluster sampling techniques. The actual procedure is described in detail in Etudes Statistiques et Econométriques , N.I.S. n° 7, 1964. See also (18).

The C.O.R.E. at Louvain owns data on food commodities and income for 30 households selected among the 1587 of the original sample. We have employed these data.[1] Before making the inferences on the income coefficient we may remark that :

1. We have considered quantities as regressands. Since we had the N.I.S. records for two types of meat, namely, first quality beef (roast beef, beefsteak, entrecote) and second quality meat (ground meat and stew), we had the choice either to consider them separately or to aggregate them. We have opted for the former procedure, given that straightforward aggregation in quantities appeared as very arbitrary (the quality and the price of the cuts being extremely different). A natural way to tackle the problem of aggregation would have been to consider total expenditure on meat as a regressand. But as we are interested in prior information on γ_{22} ($= \frac{\partial x}{\partial y}$), this procedure does not work, as can be seen from the following argument. Define e as $p_1x_1 + p_2x_2$, where the subscripts represent the two qualities of meat. Then :

$$\frac{\partial e}{\partial y} = \sum_i \left(p_i \frac{\partial x_i}{\partial y} + x_i \frac{\partial p_i}{\partial y}\right) \quad ;$$

[1]This small sample of 30 households is stratified by income. It has been selected in two steps. First the 1587 households were classified in classes of income, and the frequency distribution of the classes of income was established. In a second step, households were selected randomly within each class and in such a way that the final sample would present the same frequency distribution of the classes of income as the N.I.S. sample.

and it is impossible to obtain the value of $\frac{\partial x}{\partial y}$, unless we make the very unrealistic assumption that $\frac{\partial x_i}{\partial p_i} = 0$.

We shall be concerned, therefore, with regressing a given type of meat on income. The most appropriate type for our purpose is first quality beef, since we may reasonably expect that the total number of slaughtered animals is proportional to the supply of first quality beef. The other cuts are by-products and they will be sold at the price required to exhaust the supply.

As regressors, we have considered almost exclusively the income variable, although information on social factors was available in the N.I.S. records.[1] The income variable covers wages and fringe benefits, family allowances, old age and invalidity pensions, compensations for unemployement and sickness, and property income.

2. The quantity and income variables are the totals for the period of the enquiry.

3. We accounted for family size by considering consumption and income per capita i.e. we considered the quantity, expenditure and income variables divided by the number of persons in the household. This is a natural procedure, given that our macroeconomic variables in 1.3 are

---

[1] In regressions where we used social factors variables, their influence on the income elasticity (and the income coefficient) was negligible. The same conclusion was reached by Wold and Jureen (32) p.259.

expressed on a per capita basis. But this procedure has a consequence that can be fairly important in a cross-section study. If we assume that for all individuals the residual variance of consumption is equal, then the residual variance of the per capita consumption in the household is proportional to the reciprocal of the number of persons in the household.[1]

4. Equations of the form $x = \alpha + \beta y$ have been adjusted to the data, where x is the quantity consumed of beef, per capita, and y is the income per capita.

5. One third of the sample indicated above has been discarded as outliers. About half of the discarded households did not consume beef at all, or the consumed amounts were negligible, in spite of medium-sized incomes. The other outliers were either households with very high consumption and relatively low income or relatively low consumption and very high income. All attempts to find a systematic explanation of a socioeconomic character on those outliers have failed.

The equations have been fitted using least-squares and generalized least squares procedures. For the latter procedure we have used a diagonal covariance matrix with elements inversely proportional to the

---

[1] This is a well known consequence of an averaging process.

number of persons in the household.[1]

The results of the regressions and the implied elasticities at the sample means are given below. Remark that those results have been obtained with 20 observations.

The least squares fitting yields :

$$(1.18) \qquad x = \underset{(3.506)}{-3.274} + \underset{(0.000104)}{0.000375}y \qquad R^2 = 0.41$$

with implied elasticity :

$$(1.19) \qquad \eta_{xy} = \underset{(0.3813)}{1.3698} ,$$

whereas the generalized least-squares fitting yields :

$$(1.20) \qquad x = \underset{(3.532)}{-5.007} + \underset{(0.000107)}{0.000415}y \qquad R^2 = 0.31$$

with implied elasticity :

$$(1.21) \qquad \eta_{xy} = \underset{(0.3897)}{1.5190}$$

The results in 1.18 and 1.20 differ very slightly. We have also plotted the absolute values of the residuals, resulting from the estimations in 1.18 and 1.20, against household size in an attempt to compare

---

[1]This takes into account the heteroscedasticity of disturbances pointed out in 3.

the homoscedasticity in both models. An inspection of the scatter diagrams did not reveal any difference between the two approaches. However, given the theorical reasons in point 3 above, we shall prefer to use the results in 1.20 as a basis for our prior information.

The results in 1.18 and 1.20 are sensible. The elasticity of 1.519 in 1.21 is very similar to the elasticities obtained with time series data in 1.4 and 1.5. In a way, the value of 1.519 appears as too good, and subject to caution since it has often been remarked that income elasticities estimated from cross-section data are lower than estimates from time-series.

1.5. The complete specification of the prior density function

The prior density function $f^{o}(\beta_{12},\beta_{22},\gamma_{11},\gamma_{22})$ can now be specified completely. Our prior density function will be of the form 2.1 of Part I and the information obtained in section 1.4 above will be incorporated by observing that :

1. Information on $\delta_1' = (\beta_{12},\gamma_{11})$ (i.e. the coefficients of the supply equation) and information on $\delta_2' = (\beta_{22},\gamma_{22})$ (the coefficients of the demand equation) come from independent sources.

Indeed, information on the cattle stock coefficient comes from a regression estimated with a set of cross-section data independent of the set that has been used in the regression that gives the estimate of the income coefficient. Now, given our assumptions on the data, the estimates $\hat{\gamma}_{11}$ and $\hat{\gamma}_{22}$ in 1.16 and 1.20 are independent Normally distributed with means γ_{11}, γ_{22} and variances $V(\hat{\gamma}_{11})$ and $V(\hat{\gamma}_{22})$ depending on the variances σ_{*11} and σ_{*22} of the processes generating the cross-section data. Interchanging $\hat{\gamma}_{11}$, $\hat{\gamma}_{22}$ with γ_{11} and γ_{22}, respectively and integrating out the nuisance parameters σ_{*11} and σ_{*22}, we obtain a prior unconditional density on γ_{11} and γ_{22} that takes the form of a product of Student density functions. Furthermore, remember that we remain non-informative on the price coefficients of both equations.

Now, assume that the exponent of both Student densities is equal, say $\theta+3$. Our specification will be achieved by specifying a Student multivariate density function on β_{12} and γ_{11} (β_{22} and γ_{22} respectively) in such a way that the moments of γ_{11} (respectively γ_{22}) are as given above, and the variance of β_{12} (respectively β_{22}) is arbitrarily large. The resulting density on (δ_1, δ_2) will take the form of a product of two Student multivariate densities with parameters $(\delta_1^o, M_{11}^o, s_{11}^o, \theta+1)$ and $(\delta_2^o, M_{22}^o, s_{22}^o, \theta+1)$ respectively.

This density on (δ_1, δ_2) is reconciled with the density in 2.4 of Part I, by setting M_{12}^o, s_{12}^o and α equal to zero. We are thus able to use the developments set forth in section 2.4 of Part I. But before proceeding we need to write explicitely the first two moments of δ_1 and δ_2, since they provide a starting point for our analysis.

The first two moments of the first density are set as follows :

$$(1.22) \qquad E[\delta_1] = \delta_1^o = \begin{bmatrix} 0 \\ 109.687 \end{bmatrix}$$

$$(1.23) \qquad V[\delta_1] = \frac{s_{11}^o}{\theta-1} M_{11}^{o-1} = \begin{bmatrix} g_{11} & g_{12} \\ g_{12} & (12.459)^2 \end{bmatrix}$$

$$= \begin{bmatrix} g_{11} & g_{12} \\ g_{12} & 155.2 \end{bmatrix}$$

As we want to be non-informative on the price coefficient, its variance is set at an arbitrary large value g_{11}. The covariance is also set arbitrarily, provided that the covariance matrix $V(\delta_1)$ remains positive-definite symmetric.

For the second density, we can state similarly our prior information on the first two unconditional moments of δ_2. The first two moments will be set at :

$$(1.24) \qquad E[\delta_2] = \delta_2^o = \begin{bmatrix} 0 \\ 0.000415 \end{bmatrix}$$

$$(1.25) \qquad V[\delta_2] = \frac{s_{22}^o}{\theta-1} M_{22}^{o-1} = \begin{bmatrix} g_{11} & g_{12} \\ g_{12} & (0.0001066)^2 \end{bmatrix}$$

$$= \begin{bmatrix} g_{11} & g_{12} \\ g_{12} & 11360 \times 10^{-12} \end{bmatrix}$$

where g_{11}, again, is fixed at an arbitrarily large (but finite) value, g_{12} is also arbitrary, but $V(\delta_2)$ should remain positive-definite symmetric.

Summing up, the marginal prior information on δ_1 and δ_2 takes on the form of a product of t-multivariate densities. The moments of these densities provide a starting point for a complete specification of the prior density function, along the lines of the discussion set forth in section 2.4 of Part I.

2. The joint density of (δ_i, σ_{ii}), i = 1, 2 is assumed to be of the normal-inverted-gamma 1-type with parameter set $(\delta_i^o, M_{ii}^o, \frac{s_{ii}^o}{1-\rho^2}, \frac{\theta+1}{2})$. The two densities are independent conditionally on ρ^2.

The density of σ_{ii}, marginal with respect to δ_i but conditional with respect to ρ^2, is inverted-gamma 1 with parameter set $(\frac{s_{ii}^o}{1-\rho^2}, \frac{\theta+1}{2})$ and its first two moments are[1]:

$$(1.26) \qquad E[\sigma_{ii}] = (1-\rho^2)^{-1} \frac{s_{ii}^o}{\frac{1}{2}(\theta-1)} \qquad \theta > 1$$

$$(1.27) \qquad V[\sigma_{ii}] = \frac{(1-\rho^2)^{-1}}{2} \frac{s_{ii}^o}{(\frac{1}{2}(\theta-1))^2(\theta-3)} \qquad \theta > 3 \ .$$

---

[1] See Raiffa and Schlaiffer (22) p. 227.

The marginal density of ρ^2 is a beta density with parameters $p = 1/2$ and $q = \frac{1}{2}(\theta+2)$.

In principle we would like to be non-informative on ρ^2 (or more conveniently on ρ), and this is obtained by means of θ. The density on ρ is $f(\rho) \propto (1-\rho^2)^{\theta/2}$; with high values of θ, the density on ρ becomes more and more concentrated around 0, thus forcing us to expect an a priori low covariance between the disturbances of the two equations.[1] For selected values of θ the density $f(\rho) \propto (1-\rho^2)^{\theta/2}$ has the shapes shown in fig. 1.

Remark that even with a value as low as $\theta = 2$, the density of ρ is already concentrated around 0. The probability of $|\rho|$ being smaller than or equal to 0.35 is 0.50. The moments are $E[\rho] = 0$, $V[\rho] = \frac{1}{5}$, and $\Pr\{|\rho| \leq 2V^{1/2}(\rho)\} = 0.492$.

Unhappily, values of θ smaller or equal to 1 make us non-informative on δ_1 and δ_2 in our model. Indeed, finite variances on δ_1 and δ_2 will be available only if $\theta > 1$.

It is interesting also to note that $V(\sigma_{ii})$ does not exist for $\theta \leq 3$ as can be seen in 1.27.

3. The parameters M^o_{ii} and s^o_{ii} will be fixed proceeding as follows :
We have from 1.23 and 1.25 that the covariance matrices of δ_1 and δ_2

---

[1]For a discussion of this and related points, see Drèze and Morales (12)

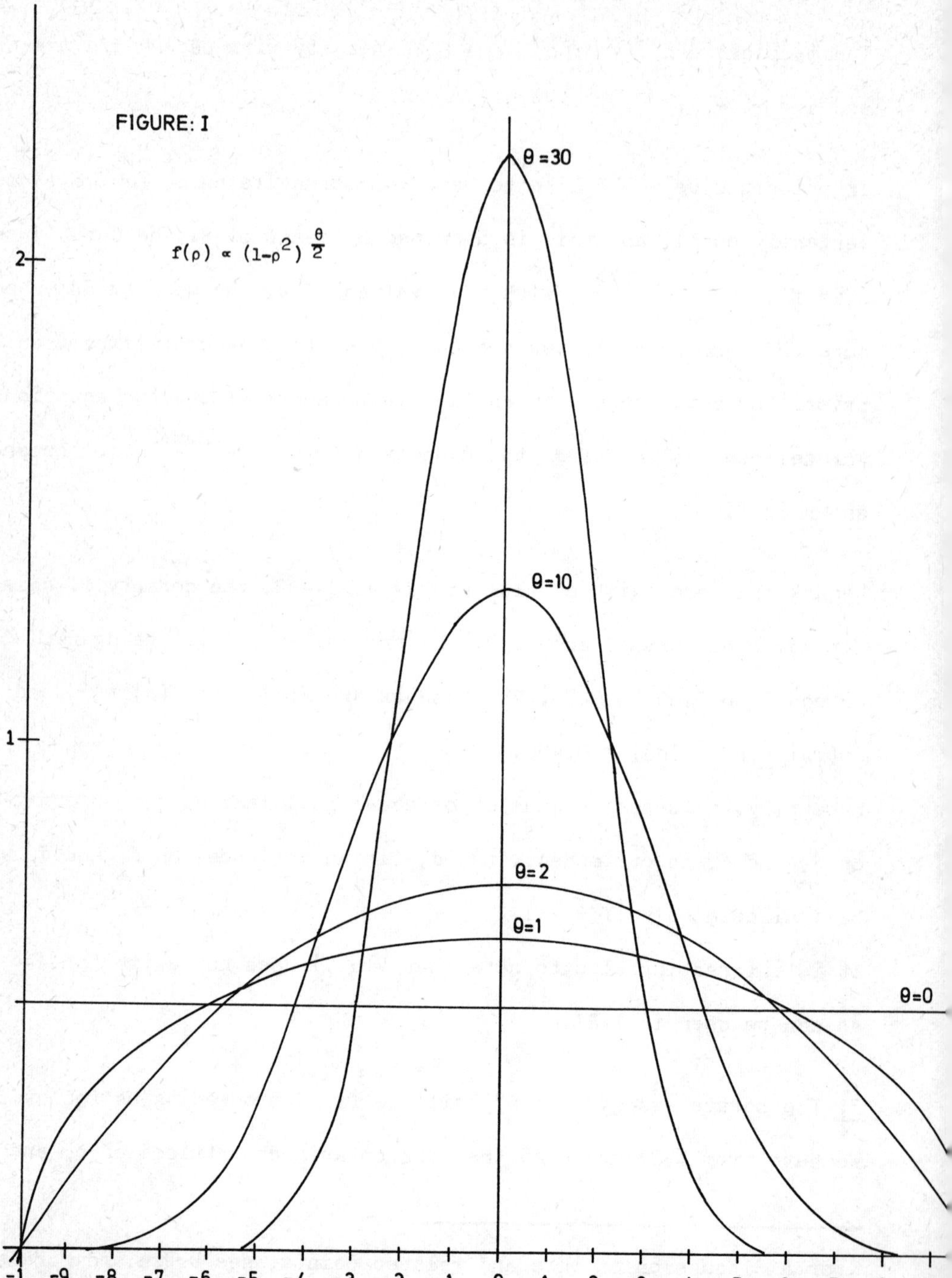
FIGURE: I
$f(\rho) \propto (1-\rho^2)^{\frac{\theta}{2}}$
θ=30
θ=10
θ=2
θ=1
θ=0
2
1
-1 -.9 -.8 -.7 -.6 -.5 -.4 -.3 -.2 -.1 0 .1 .2 .3 .4 .5 .6 .7 .8 .9

are written as :

$$V[\delta_1] = \frac{s^o_{11}}{\theta-1} M^{o-1}_{11} = \begin{bmatrix} g_{11} & g_{12} \\ g_{21} & 155 \end{bmatrix}$$

$$V[\delta_2] = \frac{s^o_{22}}{\theta-1} M^{o-1}_{22} = \begin{bmatrix} g_{11} & g_{12} \\ g_{21} & 11.360 \times 10^{-9} \end{bmatrix}$$

Therefore, the matrices M^o_{11} and M^o_{22} are written as :

$$M^o_{11} = \frac{s^o_{11}}{\theta-1} \begin{bmatrix} \dfrac{1}{g_{11} - \dfrac{g^2_{12}}{155}} & -\dfrac{g_{12}}{155g_{11}-g^2_{12}} \\ -\dfrac{g_{12}}{155g_{11}-g^2_{12}} & \dfrac{1}{155(1 - \dfrac{g^2_{12}}{g_{11}})} \end{bmatrix}$$

$$M^o_{22} = \frac{s^o_{22}}{\theta-1} \begin{bmatrix} \dfrac{1}{g_{11} - \dfrac{g^2_{12}}{11.360\times10^{-9}}} & -\dfrac{g_{12}}{11.360\times10^{-9}g_{11}-g^2_{12}} \\ -\dfrac{g_{12}}{11.360\times10^{-9}g_{11}-g^2_{12}} & \dfrac{1}{11.360\times10^{-9}(1 - \dfrac{g^2_{12}}{g_{11}})} \end{bmatrix}$$

With g_{11} (the variance of the price coefficient in the supply and demand equations) large, M^o_{11} and M^o_{22} can be written :

$$(1.28) \qquad M^{o}_{11} = \frac{s^{o}_{11}}{\theta-1} \begin{bmatrix} \varepsilon & \varepsilon' \\ \varepsilon' & \frac{1}{155} \end{bmatrix}$$

$$(1.29) \qquad M^{o}_{22} = \frac{s^{o}_{22}}{\theta-1} \begin{bmatrix} \varepsilon & \varepsilon'' \\ \varepsilon'' & \frac{10^9}{11.360} \end{bmatrix}$$

where ε is an arbitrarily small positive number, $\varepsilon' = -\frac{g_{12}}{155}$ and $\varepsilon'' = -\frac{g_{12} \times 10^9}{11.360}$.

The information furnished by the cross-section data must now be related to the information furnished by the time series.

The variance of the cattle stock coefficient in the supply equation as estimated with the cross-section data has practically the same value as the variance estimated with the time series. Thus the relativ precision in M^{o}_{11} corresponding to the cattle stock coefficient will be fixed equal to the corresponding entry in $\Xi'\Xi$, namely 0.00508. The variance of the income coefficient in the demand equation estimated with the cross-section data is little more than two times larger than the variance of the income coefficient in the time series. Thus, the prior precision of the income coefficient will be fixed at 47/100 of the value in the corresponding entry in $\Xi'\Xi$, namely 289.5×10^6.

For two values of θ, namely $\theta = 2$ and $\theta = 30$, we set the value of s^{o}_{11} (respectively s^{o}_{22}) so as to reconcile the expression on both sides of 1.28 (respectively 1.29) <u>given</u> the value of M^{o}_{11} (respectively M^{o}_{22}) as fixed in the preceeding paragraph.

For $\theta = 2$, the prior relative precision of the cattle stock coefficient is $0.00508 = s^{o}_{11}/155$; then s^{o}_{11} is set at $155 \times 0.00508 = 0.787$. Similarly, the prior relative precision of the income coefficient is $289.5 \times 10^{6} = s^{o}_{22}/(11360 \times 10^{-12})$; then s^{o}_{22} is set at $289.5 \times 10^{6} \times 11360 \times 10^{-12} = 3.288$.

For $\theta = 30$, the prior relative precision of the cattle stock coefficient and the income coefficient are $0.00508 = s^{o}_{11}/(29 \times 155)$ and $289.5 \times 10^{6} = s^{o}_{22}/(29 \times 11360 \times 10^{-12})$ respectively. Then s^{o}_{11} and s^{o}_{22} take the values 22.89 and 85.37 respectively.

The posterior analysis will also be done with $\theta = 0$ and $\theta = 1$. But, as the prior variances of δ_1 and δ_2 do not exist in these cases, we can set s^{o}_{11} and s^{o}_{22} at any arbitrary non-negative value. For convenience we will leave them at the values found with $\theta = 2$.

There is indeed a weakness in the prior specifications of M^{o}_{ii} and s^{o}_{ii} $i = 1, 2$. This appears in the specifications of the prior relative precisions that, we recall, have been chosen by comparison with the time-series data. But we are led into this way of proceeding because it is difficult to reconcile the distribution of the disturbances (or an aggregate of the disturbances) of the cross-section equations with the distribution of the disturbances in the time-series. Indeed, the time-series disturbances cannot be regarded as simple aggregates of the cross-section disturbances. This can be seen for example by considering an hypothetical economy where each household has a consumption function for beef which is linear and with a disturbance term that remains constant over time. Giving constant prices but rising incomes, the time-series demand function would give a perfect fit - but the cross-section demand function would not.

4. The final specifications of our prior density function appear therefore as :

$$(1.30) \qquad f^{o}(\delta,\Sigma^{-1}) \propto |\Sigma^{-1}|^{\frac{\theta}{2}} \exp - \frac{1}{2} \operatorname{tr}\Sigma^{-1} \left\{ \begin{bmatrix} 0.787 & 0 \\ 0 & 3.288 \end{bmatrix} \right.$$

$$+ \begin{bmatrix} \beta_{12} & \gamma_{11} - 109.7 & 0 & 0 \\ 0 & 0 & \beta_{22} & \gamma_{22} - 0.000415 \end{bmatrix}$$

$$\left. \begin{bmatrix} \varepsilon & \varepsilon' & 0 & 0 \\ \varepsilon' & 0.00508 & 0 & 0 \\ 0 & 0 & \varepsilon & \varepsilon'' \\ 0 & 0 & \varepsilon'' & 289.5\times 10^{6} \end{bmatrix} \begin{bmatrix} \beta_{12} & 0 \\ \gamma_{11} - 109.7 & 0 \\ 0 & \beta_{22} \\ 0 & \gamma_{22} - 0.000415 \end{bmatrix} \right\}$$

with $\theta = 0, 1, 2$

If $\theta = 30$, we specify our prior density function as :

$$(1.31) \qquad f^{o}(\delta,\Sigma^{-1}) \propto |\Sigma^{-1}|^{\frac{30}{2}} \exp - \frac{1}{2} tr \Sigma^{-1} \left\{ \begin{bmatrix} 22.89 & 0 \\ 0 & 85.37 \end{bmatrix} \right.$$

$$+ \begin{bmatrix} \beta_{12} & \gamma_{11} - 109.7 & 0 & 0 \\ 0 & 0 & \beta_{22} & \gamma_{22} - 0.000415 \end{bmatrix}$$

$$\left. \begin{bmatrix} \varepsilon & \varepsilon' & 0 & 0 \\ \varepsilon' & 0.00508 & 0 & 0 \\ 0 & 0 & \varepsilon & \varepsilon'' \\ 0 & 0 & \varepsilon'' & 289.5\times 10^{6} \end{bmatrix} \begin{bmatrix} \beta_{12} & 0 \\ \gamma_{11} - 109.7 & 0 \\ 0 & \beta_{22} \\ 0 & \gamma_{22} - 0.000415 \end{bmatrix} \right\}$$

2. The posterior Analysis

2.1. The posterior distributions.

In order to derive the posterior densities of δ, (and the relevant parameters of those densities), starting from the prior densities 1.30 and 1.31 and the likelihood function in 1.15 we had two alternatives, according to the developments of section 3 of Part I. First, we could combine the prior densities with the likelihood in such a way that analytical integration could be performed on the exogenous parameters γ_{11} and γ_{22}, and numerical analysis (to find the constant of integration and the moments) on β_{12}, β_{22} and Σ^{-1}. Second, we could combine the priors with the likelihood, and then, integrate out Σ^{-1} analytically. Numerical integration techniques would be used on the remaining parameters. Since the latter alternative calls for integration in a lower dimensional space, we shall follow this procedure.

The marginal density on β_{12}, β_{22}, γ_{11}, γ_{22} takes on the form 3.12 given in Part I, section 3, with parameters :$\alpha = 0$, $T = 16$, $(T-1) + \theta + m + 1 = 18 + \theta$, Δ^*, M^* and S^*. The analysis will be conducted with the values $\theta = 0, 1, 2, 30$.

Before proceeding with the posterior analysis, it is illuminating to see what are the relationships among the variables as reflected in the matrices M^*, Δ^*, S^* and especially when compared with the matrices $\Xi'\Xi$.

Ξ and Y'Y in the likelihood function 1.15. These relations will be best seen through a comparison of the correlation matrices given below in A.1, A.2, A.3.[1]

---

[1] The correlation matrix in A.1 comes from the term $A_1 M_{11.2} A_1'$ in 1.13. This expression can be rewritten as :

$$A_1 M_{11.2} A_1' = \begin{bmatrix} 1 : \delta_1' & 0 \\ 0 & 1 : \delta_2' \end{bmatrix} \begin{bmatrix} N_{11} & N_{21} \\ N_{12} & N_{22} \end{bmatrix} \begin{bmatrix} 1 : \delta_1' & 0 \\ 0 & 1 : \delta_2' \end{bmatrix}'$$

where N_{11} and N_{22} have been obtained by deleting from $M_{11.2}$ the row and the column corresponding to the income and the cattle stock variables respectively. $N_{12} = N_{21}'$ is obtained by deleting from $M_{11.2}$ the row corresponding to the cattle stock variable and the column corresponding to the income variable.
Now, we can compute the correlation matrix in A.1 by pre and post-multiplying the matrix $N = [N_{ij}]$ by a diagonal matrix containing the reciprocals of the square roots of the diagonal elements of N.

The correlation matrices in A.2 and A.3 have been computed by writing the term $Q^* = (\Delta-\Delta^*)M^*(\Delta-\Delta^*)' + S^*$, appearing in formula 3.10 of Part I, section 3 as :

$$Q^* = \begin{bmatrix} 1 : \delta_1' & 0 \\ 0 & 1 : \delta_2' \end{bmatrix} \begin{bmatrix} N_{11}^* & N_{12}^* \\ N_{21}^* & N_{22}^* \end{bmatrix} \begin{bmatrix} 1 : \delta_1' & 0 \\ 0 & 1 : \delta_2' \end{bmatrix}',$$

and pre and post-multiplying $N^* = [N_{ij}^*]$ by a diagonal matrix containing the reciprocals of the square roots of the corresponding elements of N^*.
Observe that, in order to re-arrange $Q^* = [(\Delta- \Delta^*)M^*(\Delta-\Delta^*)' + S^*]$ into

$$\begin{bmatrix} 1 : \delta_1' & 0 \\ 0 & 1 : \delta_2' \end{bmatrix} N^* \begin{bmatrix} 1 : \delta_1' & 0 \\ 0 & 1 : \delta_2' \end{bmatrix}',$$

Sample "correlation" matrix :

(A.1)

		Supply x	Supply p	Supply n	Demand x	Demand p	Demand y
Supply	x	1.00					
	p	0.61	1.00				
	n	0.95	0.53	1.00			
Demand	x	1.00	0.61	0.95	1.00		
	p	0.61	1.00	0.53	0.61	1.00	
	y	0.95	0.88	0.82	0.89	0.88	1.00

we have defined :

$$N_{11}^{*} = \left[\begin{array}{c:c} s_{11}^{*} + (\delta_{11}^{*\prime}\ \ \delta_{12}^{*\prime})M^{*}\begin{bmatrix}\delta_{11}^{*}\\ \delta_{12}^{*}\end{bmatrix} & (\delta_{11}^{*\prime}\ \ \delta_{12}^{*\prime})\begin{bmatrix}M_{11}^{*}\\ M_{21}^{*}\end{bmatrix} \\ \hdashline (M_{11}^{*}\ \ M_{12}^{*})\begin{bmatrix}\delta_{11}^{*}\\ \delta_{12}^{*}\end{bmatrix} & M_{11}^{*}\end{array}\right]$$

$$N_{22}^{*} = \left[\begin{array}{c:c} s_{22}^{*} + (\delta_{21}^{*\prime}\ \ \delta_{22}^{*\prime})M^{*}\begin{bmatrix}\delta_{21}^{*}\\ \delta_{22}^{*}\end{bmatrix} & (\delta_{21}^{*\prime}\ \ \delta_{22}^{*\prime})\begin{bmatrix}M_{12}^{*}\\ M_{22}^{*}\end{bmatrix} \\ \hdashline (M_{21}^{*}\ \ M_{22}^{*})\begin{bmatrix}\delta_{21}^{*}\\ \delta_{22}^{*}\end{bmatrix} & M_{22}^{*}\end{array}\right]$$

$$N_{12}^{*} = \left[\begin{array}{c:c} s_{12}^{*} + (\delta_{11}^{*\prime}\ \ \delta_{12}^{*\prime})M^{*}\begin{bmatrix}\delta_{21}^{*}\\ \delta_{22}^{*}\end{bmatrix} & (\delta_{11}^{*\prime}\ \ \delta_{12}^{*\prime})\begin{bmatrix}M_{12}^{*}\\ M_{22}^{*}\end{bmatrix} \\ \hdashline (M_{11}^{*}\ \ M_{12}^{*})\begin{bmatrix}\delta_{21}^{*}\\ \delta_{22}^{*}\end{bmatrix} & M_{12}^{*}\end{array}\right]$$

Prior plus sample "correlation" matrix : $\theta = 2$

(A.2)

		Supply x	Supply p	Supply n	Demand x	Demand p	Demand y
Supply	x	1.00					
	p	0.47	1.00				
	n	0.97	0.38	1.00			
Demand	x	0.61	0.49	0.54	1.00		
	p	0.47	$1-\varepsilon$	0.49	0.49	1.00	
	y	0.56	0.73	0.48	0.91	0.73	1.00

Prior plus sample "correlation" matrix : $\theta = 30$

(A.3)

		Supply x	Supply p	Supply n	Demand x	Demand p	Demand y
Supply	x	1.00					
	p	0.44	1.00				
	n	0.90	0.38	1.00			
Demand	x	0.46	0.49	0.54	1.00		
	p	0.44	$1-\varepsilon$	0.49	0.39	1.00	
	y	0.53	0.73	0.48	0.73	0.73	1.00

It is interesting to note that, as was to be expected, there is a substantial reduction in the correlations between p and n, and y and n in A.2 and A.3. The reduction in the value of the correlation between p and y is moderate. The correlations involving the dependent variables, x supply and x demand, are in general lower in A.2 and A.3 than in A.1. From the comparison between A.2 and A.3, we also remark that the correlations involving x supply and x demand are lower in A.3 than in A.2.

The marginal posterior densities $f^*(\beta_{12}|\text{data})$, $f^*(\beta_{22}|\text{data})$, $f^*(\gamma_{11}|\text{data})$, $f^*(\gamma_{22}|\text{data})$ and their respective first and second order moments have been obtained using numerical integration methods over a finite range of integration, i.e. the constants of integration, the means and the variances have been computed with a quadrature program in four dimensions using 33 points per dimension.[1] The results are summarized in table I and figures II to V.

Table I : Marginal posterior moments

	β_{12}		β_{22}		γ_{11}		γ_{22}	
	E	σ	E	σ	E	σ	E	σ
θ = 0	.170	.070	-.626	.228	109.3	8.76	.00059	.00009
θ = 1	.164	.063	-.573	.180	109.2	8.43	.00058	.00008
θ = 2	.159	.059	-.538	.149	109.2	8.14	.00057	.00007
θ = 30	.164	.060	-.728	.146	109.7	8.90	.00064	.00008

E = expectation σ = standard deviation

[1] I am greatly indebted to Jean-François Richard of C.O.R.E. at Louvain who has allowed me to use his numerical integration programs and has given me many helpful suggestions in using them.

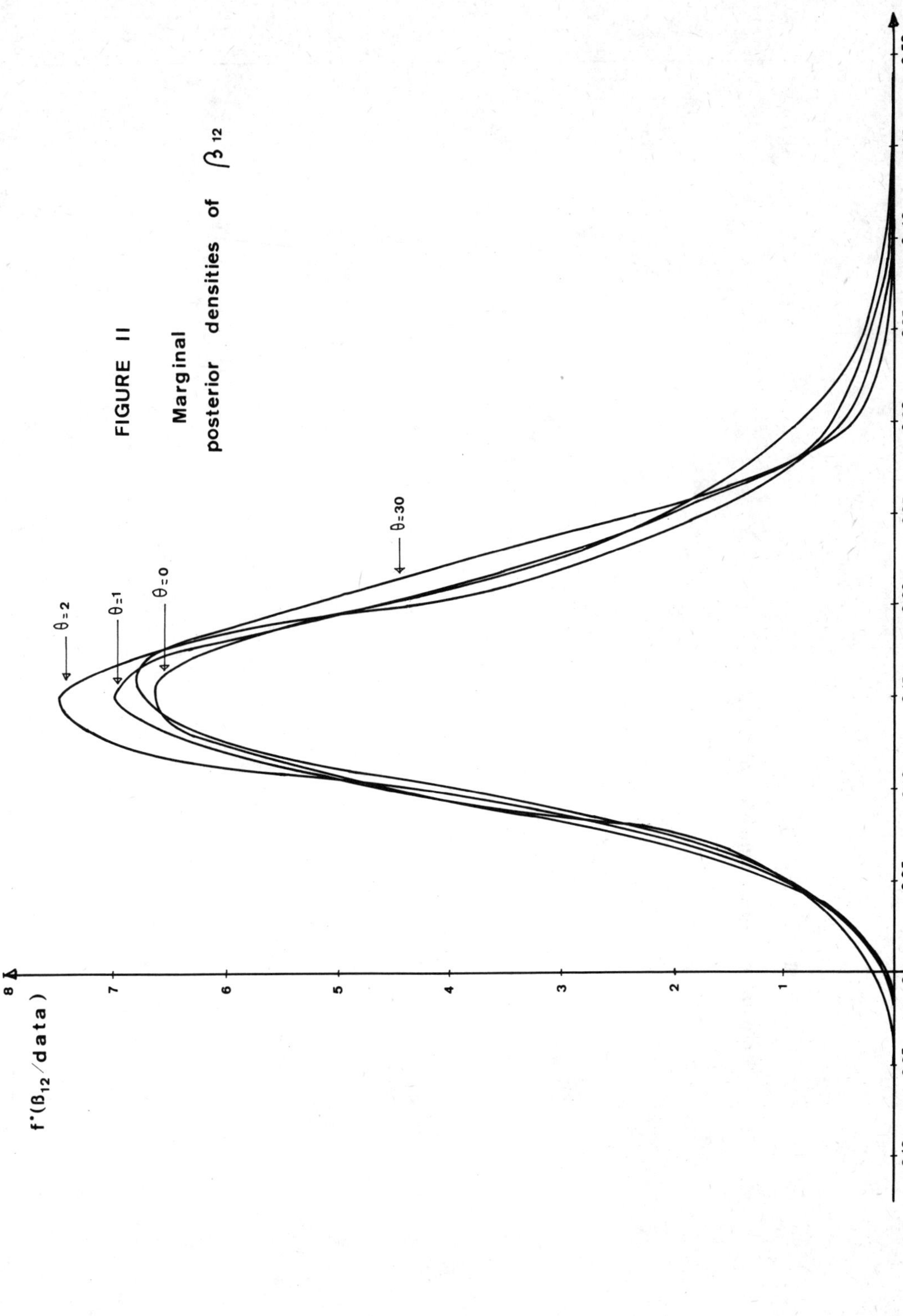

FIGURE II

Marginal posterior densities of β_{12}

FIGURE III

Marginal posterior densities of β_{22}

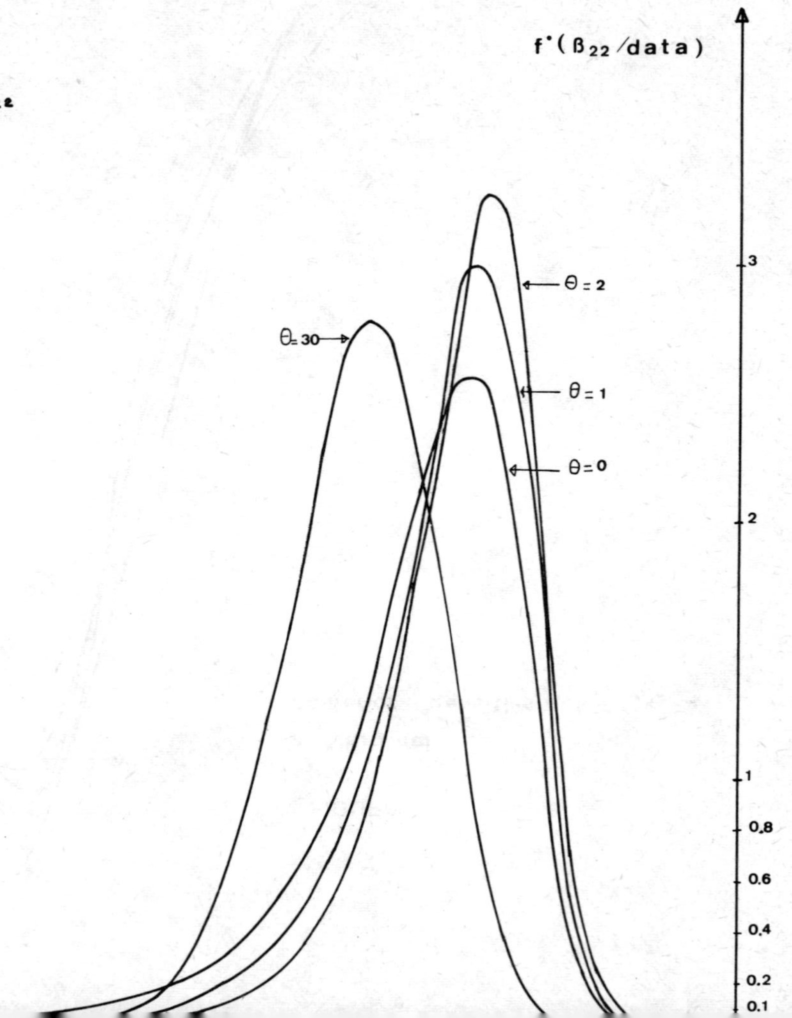

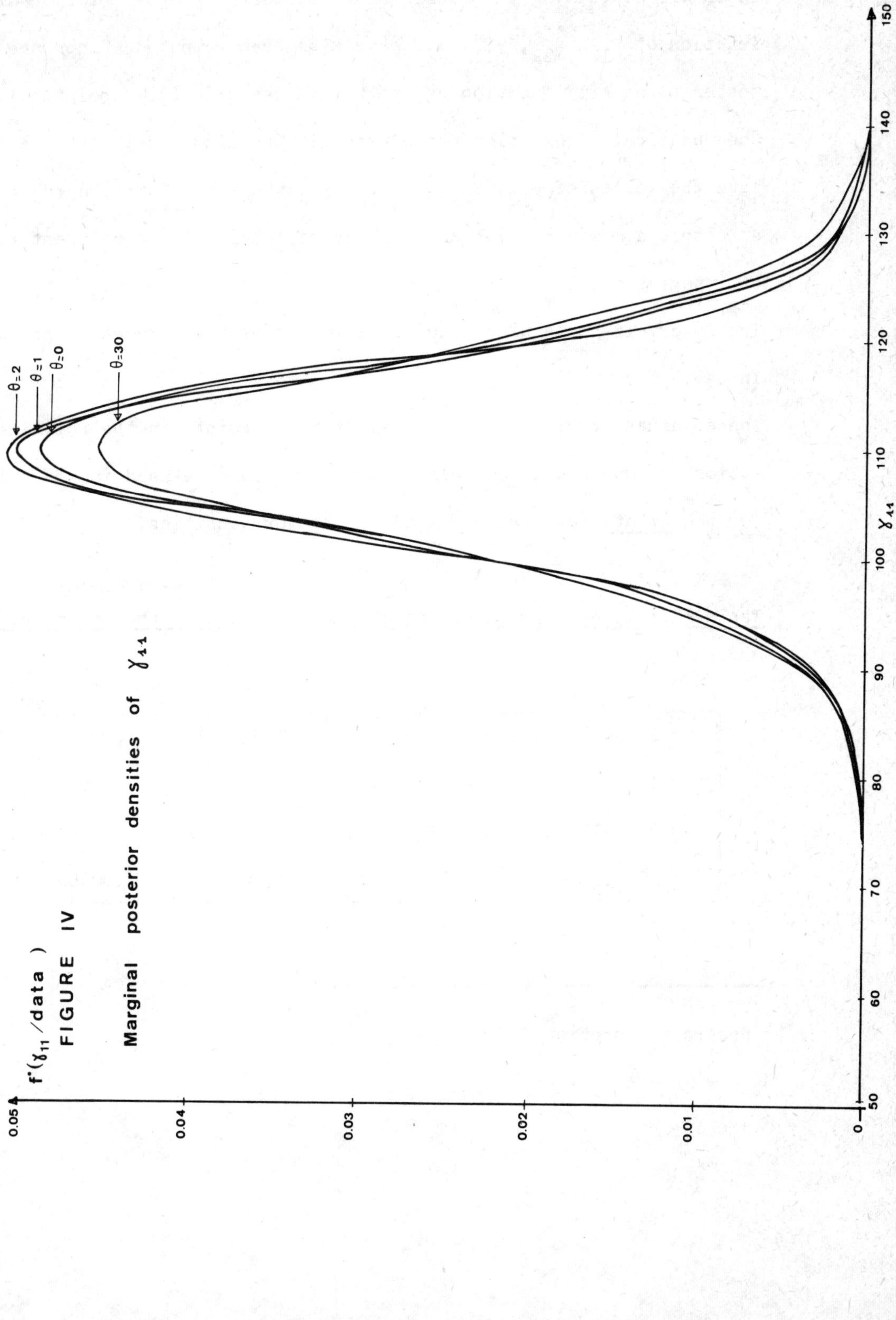

FIGURE IV

Marginal posterior densities of γ_{11}

Intervals containing the modal values of the joint posterior density function of β_{12}, β_{22}, γ_{11}, γ_{22} have also been computed. Since the posterior density function had to be evaluated in $(33)^4$ points for the numerical integration procedures, it was also possible to calculate the values of β_{12}, β_{22}, γ_{11} and γ_{22} where the function reached an absolute maximum, with an accuracy of $\pm$ 1/32 of the relevant range of integration.

The intervals containing the modal values of the parameters are given in table II.

These values provide a set of good starting points in the Full Information Maximum Likelihood algorithms that could be used in searching for the joint modal values of the posterior densities.[1]

Table II : Intervals containing the modal values of the parameters

	β_{12}	β_{22}	γ_{11}	γ_{22}
$\theta = 0$	$.14 \to .20$	$-.42 \to -.54$	$107.5 \to 112.5$	$.00054 \to .00058$
$\theta = 1$	$.13 \to .17$	$-.43 \to -.53$	$107.5 \to 112.5$	$.00052 \to .00056$
$\theta = 2$	$.13 \to .17$	$-.41 \to -.49$	$108.0 \to 112.0$	$.00051 \to .00053$
$\theta = 30$	$.14 \to .18$	$-.67 \to -.75$	$108.0 \to 112.0$	$.00062 \to .00066$

[1] See Part I section 3.2.

Following a suggestion given in Drèze and Morales[1], in order to trace the influence of θ even better, we can compare the posterior densities arising from the priors in 1.30 and 1.31 with the following specifications of the prior density function :

$$(2.1) \qquad f^{o}(\delta,\Sigma^{-1}) \propto |\Sigma|^{\frac{m+1}{2}}$$

$$(2.2) \qquad f^{o}(\delta,\Sigma^{-1}) = f^{o}(\Sigma^{-1}).f^{o}(\delta)$$

$$= |\Sigma|^{\frac{m+1}{2}} \{(\gamma_{11}-109.7)^2 \times 1.57 \times 10^{-4} + 1\}^{-\frac{44}{2}}$$

$$\{(\gamma_{22}-.0004)^2 \times 5.501 \times 10^{6} + 1\}^{-\frac{19}{2}}$$

The prior density functions in 2.1 and 2.2 combine with the likelihood function in 1.15, and the following posteriors are obtained after integrating out Σ^{-1}.

$$(2.3) \qquad f^{*}(\delta|\text{data}) \propto ||B||^{16} |A_1 M_{11.2} A_1'|^{-\frac{15}{2}}$$

$$(2.4) \qquad f^{*}(\delta|\text{data}) \propto ||B||^{16} \{(\gamma_{11}-109.7)^2 \times 1.57 \times 10^{-4} + 1\}^{-\frac{44}{2}}$$

$$\{(\gamma_{22}-.0004)^2 \times 5.501 \times 10^{6} + 1\}^{-\frac{19}{2}}$$

$$|A_1 M_{11.2} A_1'|^{-\frac{15}{2}}$$

---

[1] Drèze and Morales (12). The marginal density $f^{o}(\delta)$ in 2.2 is based on the information furnished by the cross-section data in 1.16 and 1.18. We recall that the cattle stock coefficient and the income coefficient were estimated using 45 and 20 observations respectively. The exponents 44/2 and 19/2 result from the integration of the constant term parameters. The two first moments of γ_{11} and γ_{22} are reconciled with the estimations in 1.16 and 1.18.

Table III below gives some of the relevant parameters of the marginal posteriors obtained from 2.3 and 2.4. The moments of 2.3 are not given since they do not exist. In fact, even the integral of the density 2.3 does not exist.

Table III : Modes and moments of the posterior densities obtained from prior densities 2.1 and 2.2.

1) From prior density 2.1 : Intervals containing the modes

β_{12}	β_{22}	γ_{11}	γ_{22}
.11 → .17	-.63 → -.87	107.0 → 117.0	.00070 → .0008

2) From prior density 2.2 : Intervals containing the modes

β_{12}	β_{22}	γ_{11}	γ_{22}
.13 → .17	-.54 → -.64	110.0 → 114.0	.00065 → .00069

Marginal moments

β_{12}		β_{22}		γ_{11}		γ_{22}	
E	σ	E	σ	E	σ	E	σ
.178	.07	-.623	.12	109.2	8.44	.00067	.000063

E = expectation σ = standard deviation

Table IV summarizes the results of the prior, sample and posterior analysis. The behaviour of the posterior densities to alternative prior specifications can also be compared in figures VI to IX.

Figures VI to IX exhibit the posterior densities obtained from the prior densities 1.30, with θ = 2, 1.31, with θ = 30, 2.1 and 2.2.

Table IV : Summary

	β_{12}		β_{22}		γ_{11}		γ_{22}	
	E	σ	E	σ	E	σ	E	σ
Cross-section sample	-	∞	-	∞	109.7	12.5	.00042	.00010
Time-series I.L.S.	.14	.05	-.69	.12	111.6	11.3	.00073	.00007
Posterior densities obtained from :								
1.30 θ = 0	.17	.07	-.63	.23	109.4	8.8	.00059	.00009
θ = 1	.16	.06	-.57	.18	109.2	8.4	.00057	.00008
θ = 2	.16	.06	-.54	.15	109.2	8.1	.00057	.00007
1.31 θ = 30	.16	.06	-.73	.14	109.7	8.9	.00064	.00008
2.2	.18	.07	-.62	.12	109.3	9.2	.00067	.00006

E = expectation σ = standard deviation

2.2. Comments on the results of the posterior analysis

The results of tables I to IV and figures II to IX call for the following comments :

1. Increasing the values of θ from 0 to 2 makes the densities more concentrated about their modes,as it is seen in figures II to V. This is

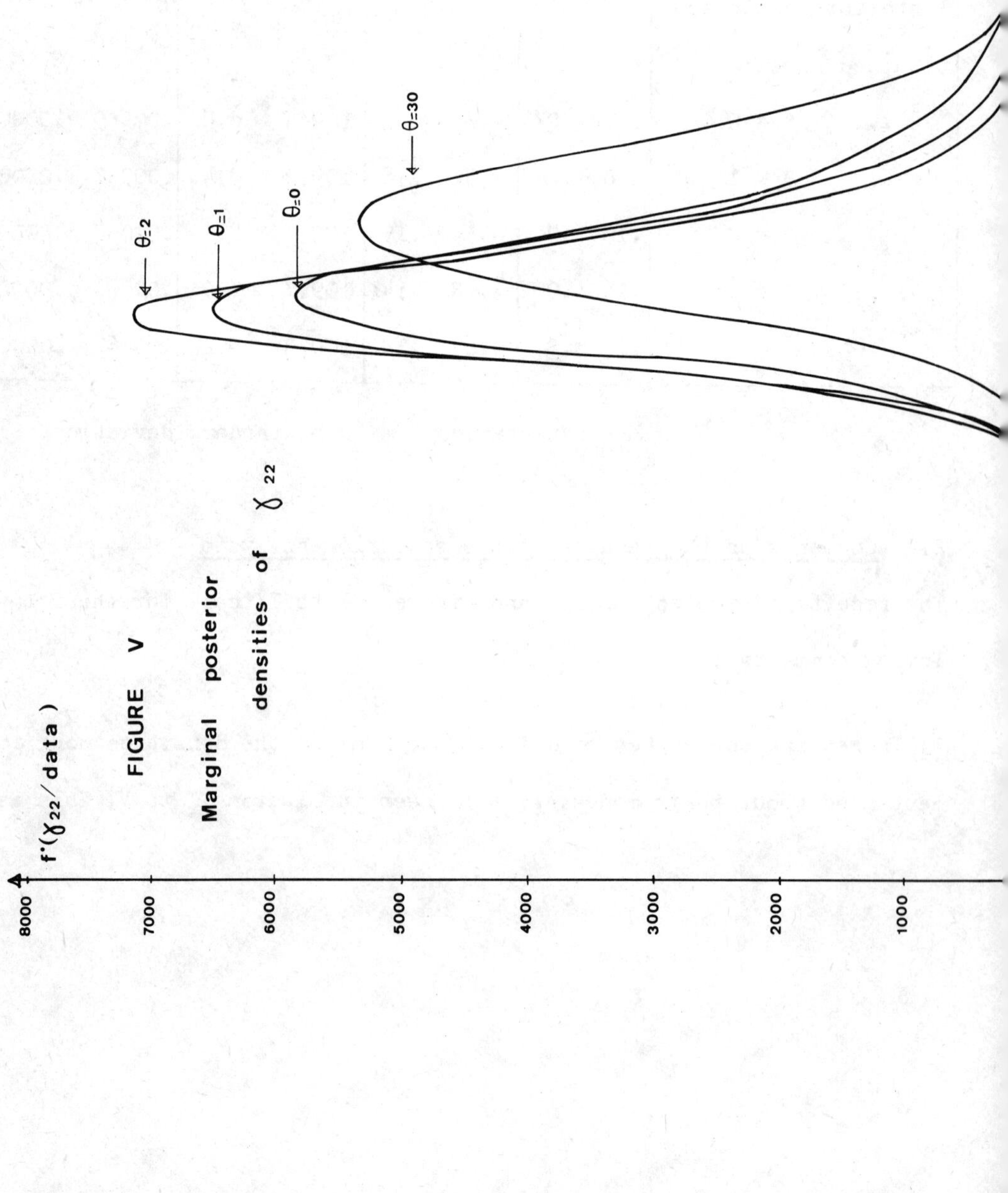

FIGURE V

Marginal posterior densities of γ_{22}

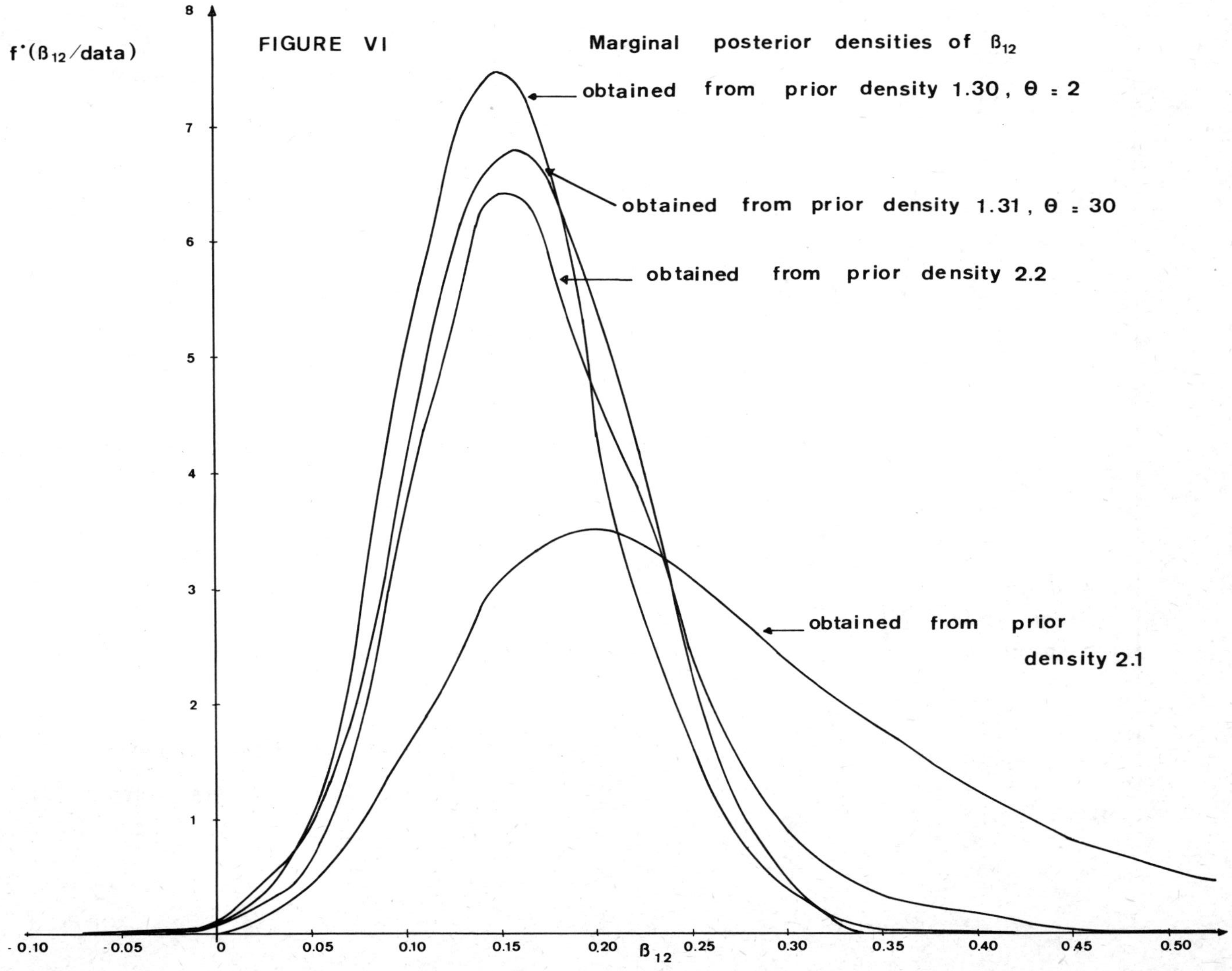

FIGURE VI

Marginal posterior densities of β_{12}

FIGURE VII

Marginal posterior densities of β_{22}

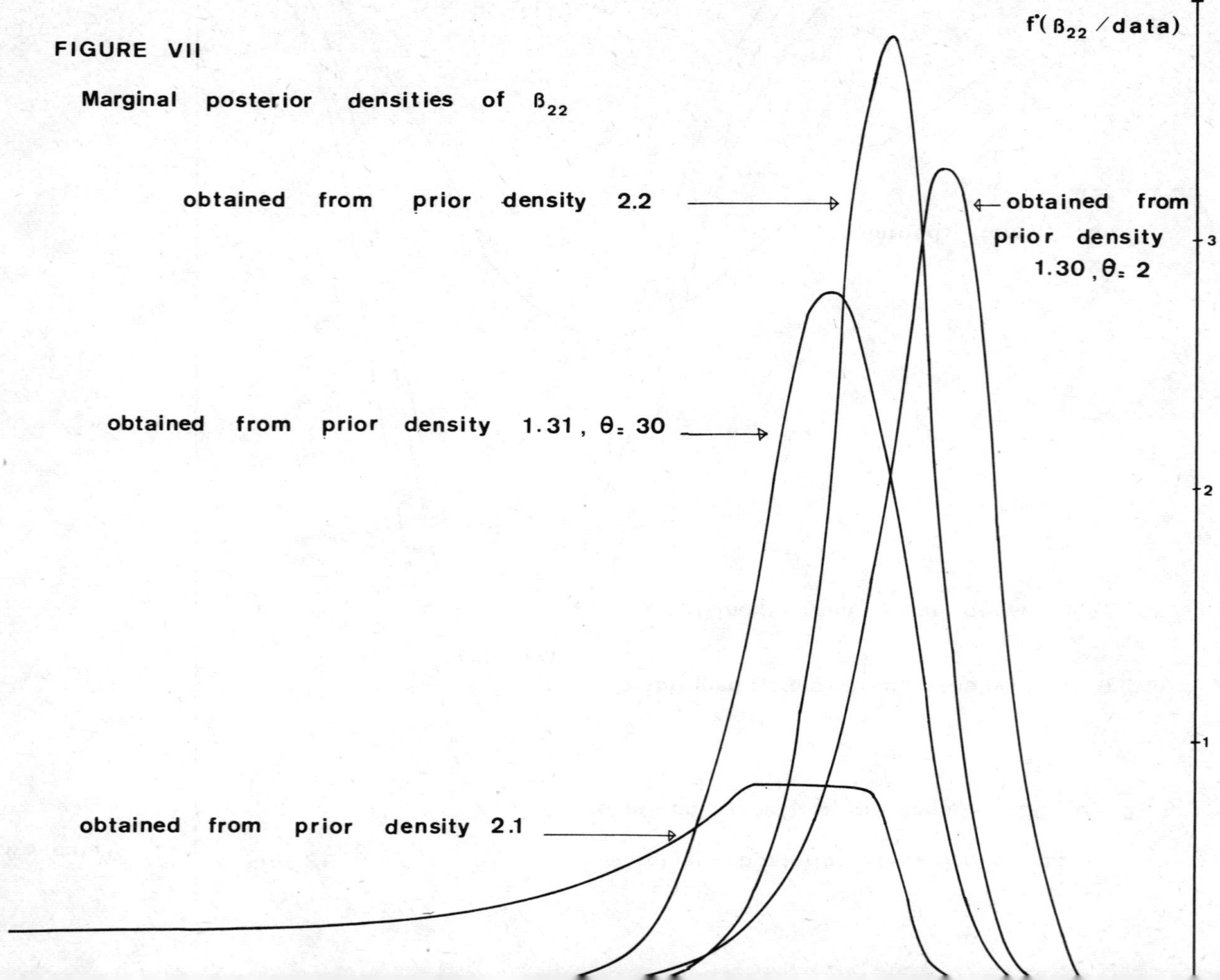

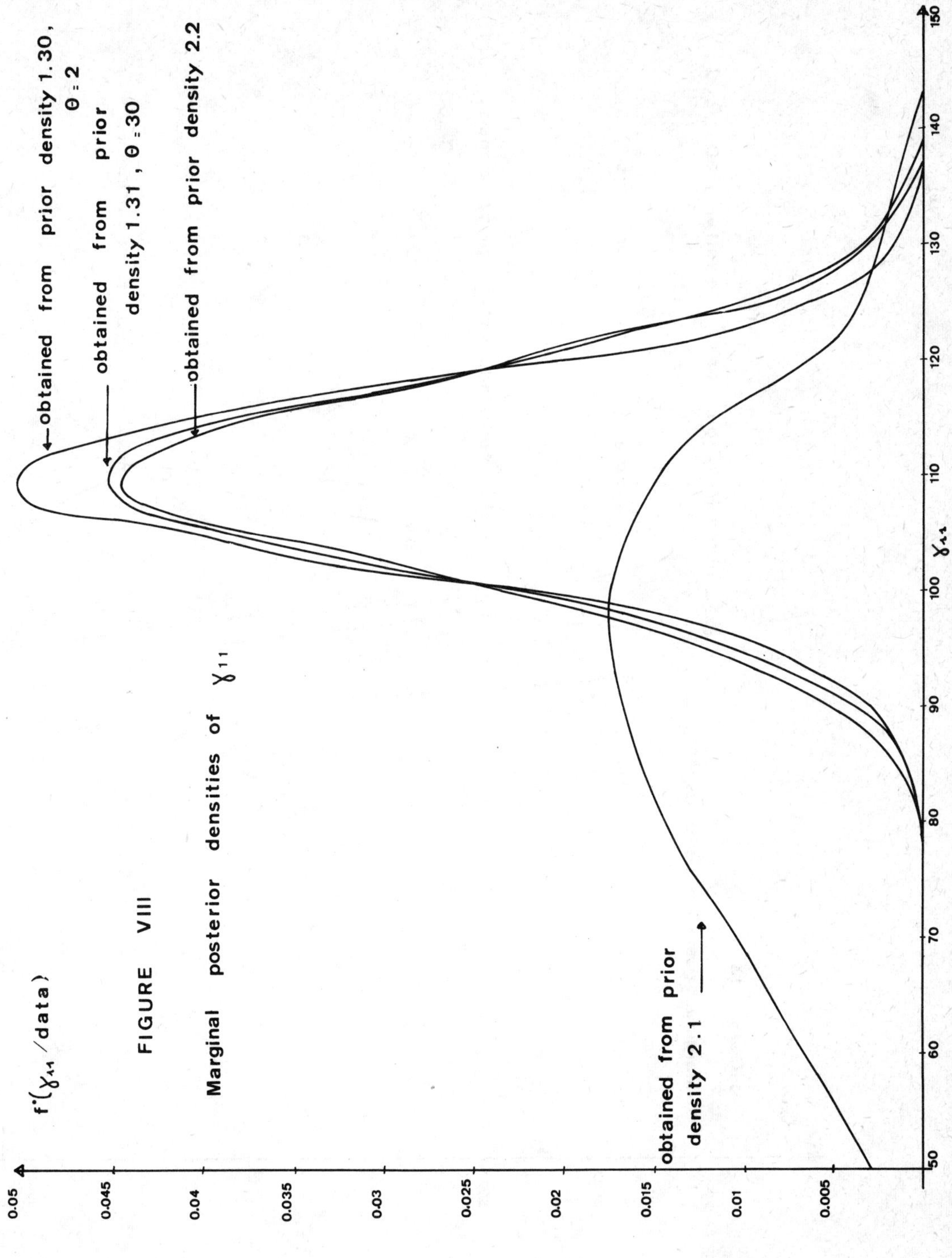

FIGURE VIII

Marginal posterior densities of γ_{11}

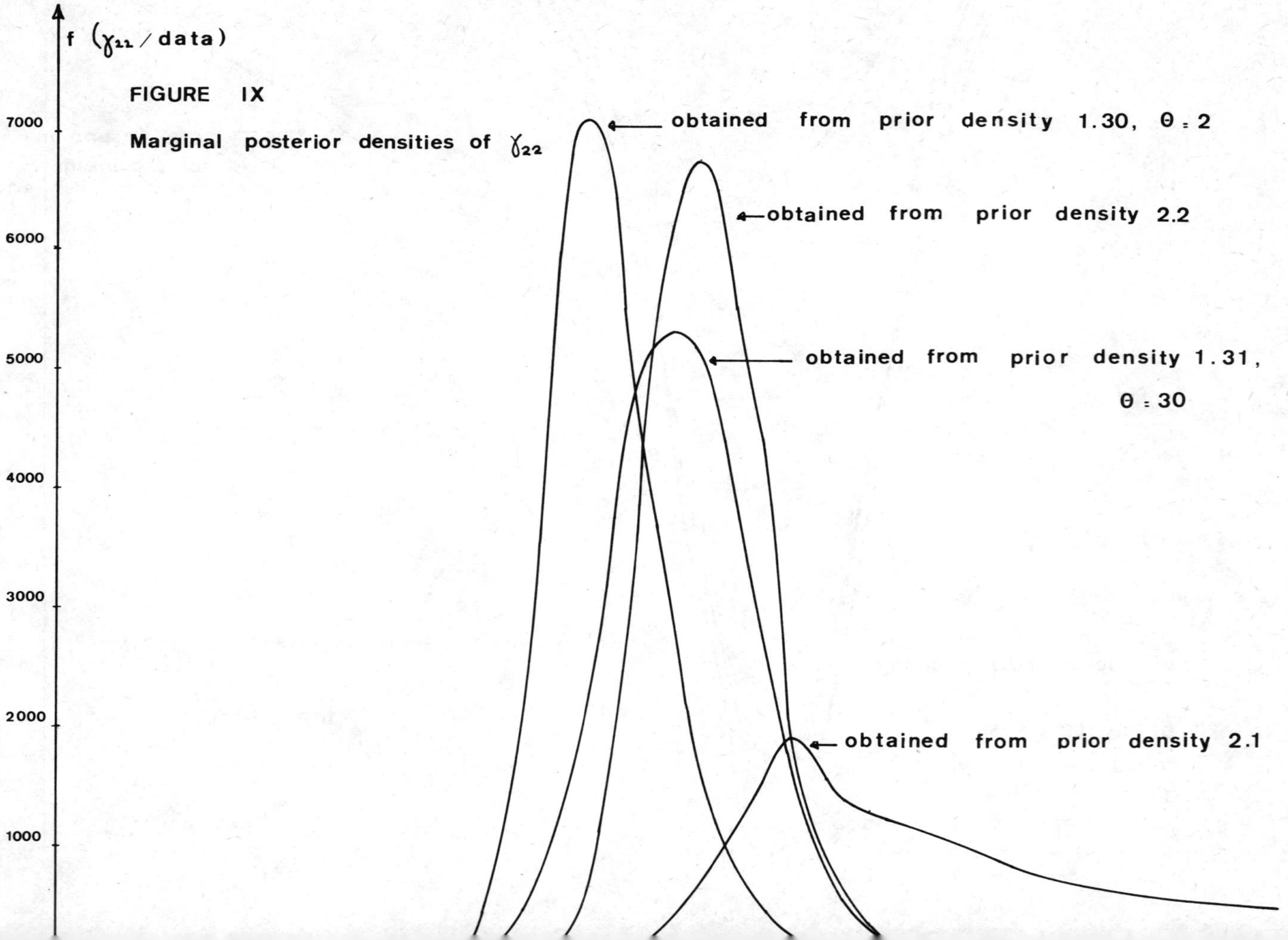

FIGURE IX

Marginal posterior densities of γ_{22}

also reflected in the decrease of the standard deviations in table I. Furthermore, for the coefficients on which we were informative, the posterior means are closer to the prior means than to the Indirect Least Squares estimates. It seems that we become fast very informative.

The results with $\theta = 30$ appear at first sight surprising; but, we recall, the values of s^o_{11} and s^o_{22} have been adjusted to θ, in order to preserve the values of the variances of γ_{11} and γ_{22}. The posterior means (with the exception of $E(\beta_{12})$) lie closer to the Indirect Least Squares estimates. The posterior standard deviations of β_{12} and β_{22} remain stationary when passing from $\theta = 2$ to $\theta = 30$. On the other hand, the posterior standard deviations of γ_{11} and γ_{22} increase rather substantially when passing from $\theta = 2$ to $\theta = 30$.

There is an interpretation to the results with $\theta = 30$, vis a vis of the results with lower values of θ. Our prior information on γ_{11} and γ_{22} is <u>modified</u> when passing from $\theta = 0, 1, 2$ to $\theta = 30$, <u>even if we preserve the variances</u> in doing the transformation. We recall that our prior marginal densities on γ_{11} and γ_{22} are Student, with parameter set γ^o_{ii}, s^o_{ii}, the corresponding element in M^o_{ii}, $\theta + 1$ $(i=1, 2)$[1]. With this parametrization, the densities are determined uniquely by four parameters. But we cannot determine uniquely the density functions by using only the first two moments, since several different values

---

[1]See section 1.4.

of the parameters s^{o}_{11}, M^{o}_{ii} and $\theta+1$, can be reconciled with a given value of the variance.

The results with $\theta = 30$ seem to be due, at least partially, to the fact that s^{o}_{11} and s^{o}_{22} have a greater value than in the cases with $\theta = 0, 1, 2$. The influence of s^{o}_{11} and s^{o}_{22} could be verified by running experiments in which s^{o}_{11} and s^{o}_{22} are increased, and θ adjusted in such a way that the prior variances are preserved. One could then analyze the behaviour of the posterior variances.

2. The posterior marginal density functions for $\theta = 0, 1, 2$ are asymmetric . This is specially marked for the posterior densities on β_{22} and γ_{22}. We recall that we had a prior information on γ_{22} that differed substantially from the sample information. Nevertheless, the differences between the (marginal) modes and the expectations on the integration intervals are not very strong, despite the asymmetries.

The densities for $\theta=30$ become symmetric, and they have a normal-type shape. With θ large, it would appear that posterior density functions of the type 3.12 in Part I converge to a Normal density function. It would be interesting to show analytically that this indeed is the case.

3. The marginal distributions of the parameters of one equation starting from a joint distribution on the parameters of the model of the

form 3.12 of Part I with $T + \alpha = 0$ have been derived in Drèze and Morales[1], Theorem 2.1. Sufficient conditions for the existence of a proper density and of moments of order p are given in corollary 2.1 to Theorem 2.1. The condition for existence of a proper density reads for our case : $\theta > 0$, since our posterior density function has the form of a determinant $(\beta_{12}-\beta_{22})$ to the power 16 times the kernel of a row-conditional matric-variate distribution with exponent $-\frac{1}{2}(18+\theta)$ The condition for existence of moments of order p reads $\theta > p$.

Thus apparently :

- the integral of the density does not exist for $\theta = 0$
- the mean does not exist for $\theta = 1$
- the variance does not exist for $\theta = 2$.

Remark that the values of the moments given in Tables I, III and IV have been computed by integrating over finite ranges. Thus they should always be understood as truncated moments. Our computations do not allow us to check if the sufficient conditions given in the corollary mentionned above are also necessary. We need further investigations on the analytical properties of densities of the form 3.12 in Part I. We also need a better understanding of the truncation errors introduced when integrating on a finite range.

4. Concerning the results obtained from prior 2.1 compared with the

[1] Drèze and Morales (12)

results from the extended natural conjugate density 1.30. and 1.31. as given in figures VI to IX, we observe, as was to be expected, very fat tails for the former. It is interesting to see that they might not even be unimodal for β_{22} and γ_{22}. The density of β_{22} has a ceiling instead of a single mode and the density of γ_{11} has a slight jump on the right side of its main mode.

5. The results obtained from prior density 2.2. seem to indicate that prior specification 2.2. is less informative than the natural conjugate specification 1.30. with $\theta = 0, 1, 2$. This is especially seen in table IV where we remark that the difference (in absolute value) between the I.L.S. estimates and the expectations of β_{22}, γ_{11}, γ_{22} increases as θ increases from 0 to 2 while this difference is less noticeable when 2.2. replaces 1.30. With $\theta = 30$, the posterior expectations are closer to the corresponding posterior expectations obtained from 2.2.

The main lesson that we can draw from figures VI to VIII is also that the posterior densities obtained from prior density 2.2. are less concentrated near the center (and so less informative) than the posterior densities obtained from the natural conjugate density with $\theta = 2$, with the exception of the density of β_{22}. But with $\theta = 30$, the posterior density of the coefficients of the demand equation is less concentrated near the center than the density of the corresponding coefficients obtained from prior density 2.2.

6. The income coefficient seems to have a strong impact on the demand price coefficient. Unhappily we do not have either the covariance between those parameters nor the contour of the joint density function. This impact is seen when comparing fig. III and V, and the means in table I or IV. With θ passing from 0 to 2, the expectation of γ_{22} decreases, and so does the expectation of β_{22} in absolute value.

7. One of the main lessons to draw from our posterior analysis is the fact that for low values of θ, the natural conjugate approach is very informative. We mean here "informative" in the sense that the posterior expectations are closer to the prior expectations than to the I.L.S. estimates. But the degree of concentration of the posterior densities about the central values diminishes as θ decreases from 2 to 0. In choosing a prior specification one should accordingly be extremely cautious in using 1.30 with low values of θ.

The choice between prior specification 1.31. (θ=30) and prior specification 2.2. is open. Some will favor the latter in view of the fact that prior information is incorporated as such, without manipulations. On the other hand, it leads to posterior densities that are asymmetric and rather concentrated about their central values.

Conclusion.

As a conclusion we can make the following general comments :

1. A Bayesian analysis of the simultaneous equations model in structural form can be performed adequately with prior densities belonging to the extended natural conjugate, as defined by 2.1 in Part I. This analysis can be performed on models with or without prior restrictions ; the treatment of the problem differs very slightly as between those cases.

Furthermore, the posterior density resulting from an extended natural conjugate prior takes a form that is analogous to the form of the prior, and could also serve as a prior density function for the analysis of new data. This reproductivity of the natural conjugate priors is indeed a very desirable property, insofar as it facilitates the accumulation of knowledge.

2. The main difficulty in a natural (or extended natural) conjugate approach lies in the fact that both the likelihood function and the prior distribution of the regression coefficients are conditional on the same precision matrix Σ^{-1}. If we are to specify the prior density function starting, for example, from sets of cross-section data, it is difficult to assert that the cross-section data are generated by the same process as the time series observations. This point has been discussed in Part II, section 2.5, for our particular model 1.3.

A further difficulty in a natural conjugate approach arises when we try to be simultaneously non-informative on the variances of the processes and marginally informative on the regression coefficients. This very important point has been discussed at some lenght when we commented on the results of our illustrative example in section 2.2 of Part II. We have to proceed very carefully with low values of θ.

3. Natural conjugate priors, and even more extended natural conjugate priors depend on a large number of parameters. For example, for our simple model 1.3 we had to specify $\frac{4(4+1)}{2}$ parameters in M^o, 4 in S^o, θ, and α. This makes a total of 18 parameters !. Of course, many parameters were set to zero, given the nature of the data on which we based our prior information. Nevertheless, in many pratical applications the large number of parameters may represent a severe limitation.

4. The analytical treatment of the posterior densities arising from natural conjugate priors seems feasible only up to a point. Analytical integration to find the normalization constant and, say, the first two moments is for the time being impossible. If we are to resort to numerical integration we can not study large size models. But, on the other hand, two equations models can be studied with reasonable accuracy and convenience, since numerical integration has

to be performed in at most a five dimensional space. This point, we recall, has been discussed in Part I, section 3.1.

For the general case, and as a second best solution, we have shown that usual techniques developed to compute the Full Information Maximum Likelihood estimators can be easily extended so as to apply to posteriors of the form 3.12 of Part I : the computations required to obtain posterior modes are similar to those required to obtain the F.I.M.L estimators.

5. Incorporation of information coming from different sources, each pertaining to one equation, seems feasible within the context of a extended natural conjugate analysis, at least for two-equations models. However, there remain the difficulties associated with the presence of a common parameter θ in the marginal densities of σ_{11}, σ_{22} and ρ^2 (or ρ). We recall that θ appears also as a parameter in the unconditional densities of δ_1, δ_2, thus relating the judgements on δ_1 and δ_2 to the judgements on ρ.

In our illustrative example, this difficulty has taken the following specific form : we wanted our prior density to imply independent Student marginal densities on δ_1 and δ_2 ; the choice of θ resulting in a non-informative marginal density on ρ then implied degenerate gamma densities for σ_{11} and σ_{22} (conditionally on ρ) ; other choices for θ, leaving the first two moments on δ_1, δ_2 unchanged were observed to influence the posterior densities on δ_1, δ_2.

This topic clearly deserves further investigation.

6. Often, in practice, not only the information on parameters of different equations comes from different sources, but also the information on parameters of the same equation comes from different sources. In the case of regression coefficients, such independence cannot very well be represented by a multivariate Student density function. Indeed, if two coefficients have independent Student densities, their joint density fails to be Student. One then has the choice of two alternatives : 1) To set the covariance of the two parameters at zero, in a multivariate Student density ; but this does not preserve their independence. 2) To decompose the joint Student in a marginal and a conditional density which, as is well known, are both Student . We can regard the information coming from one source as conditional to a value of the parameter, and the information coming from the other source as fully unconditional. A good example is provided by the demand equation in our model 1.3, where the information on the income coefficient coming from cross section data, is typically conditional to a given price coefficient, and where information on the price coefficient could only be obtained from an entirely unrelated source. But now the difficulty lies in establishing the values of the covariances.

To conclude, the main purpose of this dissertation has been to investigate the behaviour of the natural (and the extended natural) conjugate priors in the treatment of the simultaneous regression model. We have emphasized the potentials, and, alas, the weakness of this approach. We have also attempted a practical application, with all its inherent difficulties, to a real life problem that is representative of a class of problems that are present in the everyday practice of economics and econometrics. We do not mean, of course, that the results of the empirical application are ripe enough to be used in a decision-making context. In fact, in the present state of the arts, and even taking into account developments in a near future, the goals of a Bayesian analysis of the simultaneous regression model have to be rather modest. Efforts should be concentrated for a while on a better understanding of the behaviour of the posterior distributions corresponding to different prior specifications. This means that both progress in statistical distribution theory and experience in numerical methods should preceed attempts at solving specific decision problems.

References

(1) ANDERSON, T.W. (1958) An Introduction to Multivariate Statistical Analysis. New-York, Wiley and Sons.

(2) ANDO, A. and KAUFMAN, G. (1965) "Bayesian Analysis of the Independent Multinormal Process - Neither Mean nor Precision known." Journal of the American Statistical Association, Vol 60, pp. 347-358.

(3) APOSTOL, T.M. (1957) Mathematical Analysis. London, Addison - Wesley.

(4) BELLMAN, R. (1960) Introduction to Matrix Analysis. New-York Mc Graw - Hill.

(5) CALICIS, B. (1969) "Analyse Econométrique de la Demande de Viande en Belgique : 1950-1965". Recherches Economiques de Louvain, n° 2, pp. 89-109.

(6) CHERNOFF, H. and DIVINSKY, N. (1953) "The Computation of Maximum Likelihood Estimates of Linear Structural Equations". Chapter X in W.C. HOOD and T.C. KOOPMANS, Studies in Econometric Method, New-York, Wiley and Sons, pp. 236-302.

(7) DESAYERE, W., STUYCK, H., VAN BROEKHOVEN, E., VAN HAEPEREN,J.M. (1967) Long Term Development of Supply and Demand for Agricultural Products in Belgium 1970-1975. Studiecentrum voor Ekonomisch en Sociaal Onderzoek (SESO) Antwerp.

(8) DHRYMES, Ph.(1970) Econometrics.Statistical Foundations and Applications, New-York, Harper and Row.

(9) DICKEY, J.M. (1967) "Matric-variate Generalizations of the Multivariate- t Distribution and the inverted Multivariate-t Distribution". Annals of Mathematical Statistics, 38, pp. 511-518.

(10) DREZE, J.H. (1962) "The Bayesian Approach to Simultaneous Equations Estimation". O.N.R. Research Memorandum, n° 67, Northwestern University.

(11) DREZE, J.H. (1968) "Limited Information Estimation from a Bayesian viewpoint". CORE Discussion Paper n° 6816, Louvain.

(12) DREZE, J.H. and MORALES, J.A. (1970) "Bayesian Full Information Analysis of the Simultaneous Equation Model". CORE Discussion Paper n° 7031, Louvain.

(13) FAURE, H. (1967) "Etude Econométrique de la Demande de Viande". Consommation, Annales du C.R.E.D.O.C. n°1, Paris, Dunod.

(14) FELLER, W. (1966) An Introduction to Probability Theory and its Applications.Vol II, New-York, Wiley and Sons.

(15) FISHER, F.M. (1966) The Identification Problem in Economics. New-York, Mc Graw-Hill.

(16) GOLDBERGER, A.S. (1964) Econometric Theory . New-York, Wiley and Sons.

(17) Institut d'Economie Agricole (1967-1968) Feuilles des Résultats comptables.
See also : "Aperçu des Résultats comptables moyens de 913 exploitations agricoles". Cahiers de l'I.E.A., n° 98/ EE- 13, Bruxelles.

(18) Institut National de Statistique.Royaume de Belgique. (1950-1968) Annuaires Statistiques, Bruxelles.
1963-1964-1965 "Enquête sur les Budgets de Ménages" Etudes Statistiques et Econométriques n° 5,7,9.

(19) KAUFMAN, G. (1967) "Some Bayesian Moment Formulae" CORE Discussion Paper n° 6710, Louvain.

(20) KENDALL, M.G. and STUART, A. (1958) The advanced Theory of Statistics, Vol I., London, Griffin.

(21) KOLGOMOROV, A.N. and FOMIN, S.V. (1961) Measure, Lebesgue Integrals and Hilbert Spaces. New-York, Academic Press.

(22) RAIFFA, H. and SCHLAIFFER, R. (1961) Applied Statistical Decision Theory. Harvard University Press, Cambridge.

(23) ROTHENBERG, T.J. and LEENDERS, C.T. (1962) "Efficient Estimation of Simultaneous Equation Systems" Econometrica, 32, pp 57-76.

(24) ROTHENBERG, T.J. (1963) "A Bayesian Analysis of Simultaneous Equations Systems" Report 6315 Econometric Institute, Rotterdam.

(25) ROTHENBERG, T.J. (1968) "The value of Structural Information : A Bayesian Approach" CORE Discussion paper n° 6814 - Louvain.

(26) ROTHENBERG, T.J. (1969) "Identification in Parametric Models" CORE Discussion Paper, n° 6909, Louvain.

(27) RUBBLE, W.L. (1968) "Improving the Computation of Simultaneous Stochastic Linear Equations Estimates". Agricultural Economics Report n° 116 and Econometrics Special Report n°1. Departement of Agricultural Economics, Michigan State University, East Lansing.

(28) THEIL and GOLDBERGER (1961) "On Pure and Mixed Statistical Estimation in Economics". International Economic Review, Vol 2, n° 1, pp. 65-78.

(29) TIAO, G.C. and ZELLNER, A. (1964) "On the Bayesian Estimation of Multivariate Regression" Journal of the Royal Statistical Society Series B, pp. 277-285.

(30) TOMEK, W.G. and COCHRANE, W.W. (1962) "Long-Run Demand : A Concept and Elasticity Estimates for Meats". Journal of Farm Economics, 44, pp. 721-730.

(31) WEGGE, L.L. (1969) "The Finite Sampling Distribution of Least Squares Estimators with Stochastic Regressors" Unpublished Manuscript.

(32) WOLD, H. and JUREEN, L. (1962) Demand Analysis. New-York, Wiley and Sons.

(33) ZELLNER, A., and THEIL, H. (1962) "Three stage least-squares ; Simultaneous Estimation of Simultaneous Equations". Econometrica, 30, pp. 54-78.

(34) ZELLNER, A. (1962) "An efficient Method for Estimating Seemingly Unrelated Regressions and Test for Aggregation Biais". Journal of the American Statistical Association, 60, pp. 608-616.

(35) ZELLNER, A. Introduction to Bayesian Inference in Econometrics. (forthcoming).

Lecture Notes in Operations Research and Mathematical Systems

Vol. 1: H. Bühlmann, H. Loeffel, E. Nievergelt, Einführung in die Theorie und Praxis der Entscheidung bei Unsicherheit. 2. Auflage, IV, 125 Seiten 4°. 1969. DM 12,– / US $ 3.30

Vol. 2: U. N. Bhat, A Study of the Queueing Systems M/G/1 and GI/M/1. VIII, 78 pages. 4°. 1968. DM 8,80 / US $ 2.50

Vol. 3: A. Strauss, An Introduction to Optimal Control Theory. VI, 153 pages. 4°. 1968. DM 14,– / US $ 3.90

Vol. 4: Einführung in die Methode Branch and Bound. Herausgegeben von F. Weinberg. VIII, 159 Seiten. 4°. 1968. DM 14,– / US $ 3.90

Vol. 5: L. Hyvärinen, Information Theory for Systems Engineers. VIII, 205 pages. 4°. 1968. DM 15,20 / US $ 4.20

Vol. 6: H. P. Künzi, O. Müller, E. Nievergelt, Einführungskursus in die dynamische Programmierung. IV, 103 Seiten. 4°. 1968. DM 9,– / US $ 2.50

Vol. 7: W. Popp, Einführung in die Theorie der Lagerhaltung. VI, 173 Seiten. 4°. 1968. DM 14,80 / US $ 4.10

Vol. 8: J. Teghem, J. Loris-Teghem, J. P. Lambotte, Modèles d'Attente M/G/1 et GI/M/1 à Arrivées et Services en Groupes. IV, 53 pages. 4°. 1969. DM 6,– / US $ 1.70

Vol. 9: E. Schultze, Einführung in die mathematischen Grundlagen der Informationstheorie. VI, 116 Seiten. 4°. 1969. DM 10,– / US $ 2.80

Vol. 10: D. Hochstädter, Stochastische Lagerhaltungsmodelle. VI, 269 Seiten. 4°. 1969. DM 18,– / US $ 5.00

Vol. 11/12: Mathematical Systems Theory and Economics. Edited by H. W. Kuhn and G. P. Szegö. VIII, IV, 486 pages. 4°. 1969. DM 34,– / US $ 9.40

Vol. 13: Heuristische Planungsmethoden. Herausgegeben von F. Weinberg und C. A. Zehnder. II, 93 Seiten. 4°. 1969. DM 8,– / US $ 2.20

Vol. 14: Computing Methods in Optimization Problems. Edited by A. V. Balakrishnan. V, 191 pages. 4°. 1969. DM 14,– / US $ 3.90

Vol. 15: Economic Models, Estimation and Risk Programming: Essays in Honor of Gerhard Tintner. Edited by K. A. Fox, G. V. L. Narasimham and J. K. Sengupta. VIII, 461 pages. 4°. 1969. DM 24,– / US $ 6.60

Vol. 16: H. P. Künzi und W. Oettli, Nichtlineare Optimierung: Neuere Verfahren, Bibliographie. IV, 180 Seiten. 4°. 1969. DM 12,– / US $ 3.30

Vol. 17: H. Bauer und K. Neumann, Berechnung optimaler Steuerungen, Maximumprinzip und dynamische Optimierung. VIII, 188 Seiten. 4°. 1969. DM 14,– / US $ 3.90

Vol. 18: M. Wolff, Optimale Instandhaltungspolitiken in einfachen Systemen. V, 143 Seiten. 4°. 1970. DM 12,– / US $ 3.30

Vol. 19: L. Hyvärinen, Mathematical Modeling for Industrial Processes. VI, 122 pages. 4°. 1970. DM 10,– / US $ 2.80

Vol. 20: G. Uebe, Optimale Fahrpläne. IX, 161 Seiten. 4°. 1970. DM 12,– / US $ 3.30

Vol. 21: Th. Liebling, Graphentheorie in Planungs- und Tourenproblemen am Beispiel des städtischen Straßendienstes. IX, 118 Seiten. 4°. 1970. DM 12,– / US $ 3.30

Vol. 22: W. Eichhorn, Theorie der homogenen Produktionsfunktion. VIII, 119 Seiten. 4°. 1970. DM 12,– / US $ 3.30

Vol. 23: A. Ghosal, Some Aspects of Queueing and Storage Systems. IV, 93 pages. 4°. 1970. DM 10,– / US $ 2.80

Vol. 24: Feichtinger, Lernprozesse in stochastischen Automaten. V, 66 Seiten. 4°. 1970. DM 6,– / $ 1.70

Vol. 25: R. Henn und O. Opitz, Konsum- und Produktionstheorie I. II, 124 Seiten. 4°. 1970. DM 10,– / $ 2.80

Vol. 26: D. Hochstädter und G. Uebe, Ökonometrische Methoden. XII, 250 Seiten. 4°. 1970. DM 18,– / $ 5.00

Vol. 27: I. H. Mufti, Computational Methods in Optimal Control Problems. IV, 45 pages. 4°. 1970. DM 6,– / $ 1.70

Vol. 28: Theoretical Approaches to Non-Numerical Problem Solving. Edited by R. B. Banerji and M. D. Mesarovic. VI, 466 pages. 4°. 1970. DM 24,– / $ 6.60

Vol. 29: S. E. Elmaghraby, Some Network Models in Management Science. III, 177 pages. 4°. 1970. DM 16,– / $ 4.40

Vol. 30: H. Noltemeier, Sensitivitätsanalyse bei diskreten linearen Optimierungsproblemen. VI, 102 Seiten. 4°. 1970. DM 10,– / $ 2.80

Vol. 31: M. Kühlmeyer, Die nichtzentrale t-Verteilung. II, 106 Seiten. 4°. 1970. DM 10,– / $ 2.80

Vol. 32: F. Bartholomes und G. Hotz, Homomorphismen und Reduktionen linearer Sprachen. XII, 143 Seiten. 4°. 1970. DM 14,– / $ 3.90

Vol. 33: K. Hinderer, Foundations of Non-stationary Dynamic Programming with Discrete Time Parameter. VI, 160 pages. 4°. 1970. DM 16,– / $ 4.40

Vol. 34: H. Störmer, Semi-Markoff-Prozesse mit endlich vielen Zuständen. Theorie und Anwendungen. VII, 128 Seiten. 4°. 1970. DM 12,– / $ 3.30

Vol. 35: F. Ferschl, Markovketten. VI, 168 Seiten. 4°. 1970. DM 14,– / $ 3.90

Vol. 36: M. P. J. Magill, On a General Economic Theory of Motion. VI, 95 pages. 4°. 1970. DM 10,– / $ 2.80

Vol. 37: H. Müller-Merbach, On Round-Off Errors in Linear Programming. VI, 48 pages. 4°. 1970. DM 10,– / $ 2.80

Vol. 38: Statistische Methoden I, herausgegeben von E. Walter. VIII. 338 Seiten. 4°. 1970. DM 22,– / $ 6.10

Vol. 39: Statistische Methoden II, herausgegeben von E. Walter. IV, 155 Seiten. 4°. 1970. DM 14,– / $ 3.90

Vol. 40: H. Drygas, The Coordinate-Free Approach to Gauss-Markov Estimation. VIII, 113 pages. 4°. 1970. DM 12,– / $ 3.30

Vol. 41: U. Ueing, Zwei Lösungsmethoden für nichtkonvexe Programmierungsprobleme. IV, 92 Seiten. 4°. 1971. DM 16,– / $ 4.40

Vol. 42: A.V. Balakrishnan, Introduction to Optimization Theory in a Hilbert Space. IV, 153 pages. 4°, 1971. DM 16,– / $ 4.40

Vol. 43: J. A. Morales, Bayesian Full Information Structural Analysis. VI, 154 pages. 4°, 1971. DM 16,– / $ 4.40

Vol. 44: G. Feichtinger, Stochastische Modelle demographischer Prozesse. XIII, 404 pages. 4°, 1971. DM 28,– / $ 7.70

Vol. 45: K. Wendler, Hauptaustauschschritte (Principal Pivoting). II, 65 pages. 4°, 1971. DM 16,– / $ 4.40

Vol. 46: C. Boucher, Leçons sur la théorie des automates mathématiques. VIII, 193 pages. 4°, 1971. DM 18,– / $ 5.00